WiMAX Modeling: Techniques and Applications

T0172147

Bhavin S. Sedani • Komal R. Borisagar
Rohit M. Thanki

WiMAX Modeling: Techniques and Applications

 Springer

Bhavin S. Sedani
E. C. Department
L. D. Engineering College
Ahmedabad, India

Rohit M. Thanki (iD)
Faculty of Technology and Engineering
C. U. Shah University
Wadhwan City, Gujarat, India

Komal R. Borisagar
Mobile Computing and Networking
Technology
Graduate School of Engineering and
Technology
Gujarat Technological University
Ahmedabad, India

ISBN 978-3-030-22462-2   ISBN 978-3-030-22460-8   (eBook)
https://doi.org/10.1007/978-3-030-22460-8

This Springer imprint is published by the registered company Springer Nature Switzerland AG
The registered company address is: Gewerbestrasse 11, 6330 Cham, Switzerland

Preface

The world without wires is constantly popular and it is developing at a gigantic speed on account of the most significant element of being exceptionally portable in nature. Wireless gadgets, for example, cell phones have been accomplishing increasingly more fascination principally on account of their mobility. Despite the fact that at first, the early telephones were perfect with voice applications just, as an inventive information benefits as far as content have been included and the most recent errand is toward mixed media administrations, for example, move of recordings and pictures have exceptionally included into it. The utilization and implementation of an innovative data services in real life are continuously growing. The most significant property of any fast growing wireless communication system is to give the perpetual progression high information rate, high capacity and less time utilization, which is turning into the neediest necessity for the future age wireless frameworks. However the primary obstruction that is experienced by the emerging wireless networks is to develop the wireless framework with sufficiently high speed that requires a high amount of bandwidth and lower Bit Error Rate. The other major occurrences of wireless communication that makes the problem challenging and fascinating are fading and interference. To beat the above expressed issues, possibly one has to rely on the wired system which isn't anything but difficult to deploy in remote rustic territories in view of absence of versatility, or needs to build up the wireless system with adequately requires a high measure of data transfer capacity.

To overcome the above stated problems, either one has to rely on the wired network which is not easy to deploy in remote rural areas because of lack of mobility, or has to develop the wireless network with sufficiently high speed that requires a high amount of bandwidth. Mobile broadband wireless access offers an elastic and gainful solution to these problems. In recent years, the WiMAX (Worldwide Interoperability for Microwave Access) standard has materialized to harmonies the wide variety of different BWA (Broadband Wireless Access) technologies. To fulfill every aspect of the modern wireless communication systems, i.e., higher bit rate, lower bit error rate, and greater capacity, the most emerging networking standard WiMAX is the best solution. WiMAX is the most promising wireless networking standard having the unique features of 50 km of coverage range as well as throughput up to 70 Mbps to cope up with the current requirement. Alongside the higher information rate, the above expressed two significant occurrences of wireless communication for example fading and interference can be limited by means of the implementation of antenna diversity schemes in the WiMAX framework as the WiMAX framework is supporting advanced antenna systems.

Be that as it may, in the present situation, the WiMAX framework has been practically implemented with the traditional single transmitter-receiver system. For the real-time transmission of image and speech signals rather than just data with higher quality and faster data rate, under any kind of environments for example, AWGN, Rayleigh, or Rician, it is important to refresh the acknowledgment of WiMAX framework. This is the requirement of antenna diversity schemes at the transmitter and/or at the receiver side to be implemented in a WiMAX system along with the Alamouti coding scheme. As a whole to provide endless mobility with ultimate capacity and reduced bit error rate, the most promising fourth-generation technique is WiMAX system with the implementation of antenna diversity techniques such as MIMO technique along with the most sophisticated space-time coding, i.e., Alamouti coding for real-time data or image or speech transfer. This book gives fundamental information about the modeling of WiMAX framework. Also the various techniques such as antenna diversity algorithms and Alamouti coding technique are discussed for WiMAX frame work. The WiMAX modeling utilizing these strategies is as a rule delightfully discussed and presented. The presentation of WiMAX framework is tested utilizing different sorts of information, for example, speech and image signal transmission under different kinds of channel environments.

Ahmedabad, Gujarat, India   Bhavin S. Sedani
Rajkot, Gujarat, India   Komal R. Borisagar
                                                               Rohit M. Thanki

Acknowledgments

It has been a fun writing this book. My task has been easier, and the final version of the book is considerably better because of the help we have received. Acknowledging that help is itself a pleasure. We would extend many thanks to all persons whose assistance helped achieve the final version of this book. This book is a Ph.D. research work and extension work of Dr. Bhavin Sedani, submitted to the Department of Electronics and Communication Engineering, Shri Jagdishprasad Jhabarmal Tibrewala University (JJTU), Jhunjhunu, Rajasthan in 2012. The authors are indebted to numerous colleagues for valuable suggestions during the entire period of the manuscript preparation. We would also like to thank the publishers at Springer, in particular Mary James, senior publishing editor/CS Springer, for their helpful guidance and encouragement during the creation of this book.

Contents

List of Figures

List of Tables

About the Authors

Bhavin S. Sedani is working as a professor in Electronics and Communication Department at L.D. College of Engineering, Ahmedabad, India. He has teaching experience of 16 years. He has presented more than 46 research papers in various international and national conferences. His 21 research papers are published in various international journals and IEEE Xplore. He has achieved best paper presentation awards seven times for his research articles. He is awarded pedagogical award for continuing efforts toward teaching-learning methodology, research, and innovation in 2017 by Gujarat Technological University and Young Researcher Award in 2018 by Integrated Intelligent Research, ISTE Professional Center, Anna University Campus, Chennai—India. He has filed Indian patent with title *Brain Machine Technology based Wireless Blazing of Traditional Indian Wick Lamp*. His area of research is wireless communication and speech processing.

Komal R. Borisagar is working as an Associate Professor in Electronics and Communication Department at Atmiya Institute of Technology and Science, Rajkot, India. She has teaching experience of 16 years. She has published 3 books, 3 book chapters, and more than 50 research papers to her credit in referred and indexed journals, conferences at international and in IEEE digital library. She has achieved best paper award five times for her research articles and presentation. Her areas of interest are wireless communication, speech processing, signal and system, and image processing.

Rohit M. Thanki received his Ph.D. in Electronics and Communication Engineering from C. U. Shah University, M.E. in Communication Engineering from G. H. Patel College of Engineering and Technology, and B.E. in Electronics and Communication Engineering from Atmiya Institute of Technology and Science, India. He has more than 3 years of experience in academic and research. He has published 9 books with Springer and 1 book with CRC press. He has published 13 book chapters in edited books which are published by Elsevier, Springer, CRC Press, and IGI Global. He has also published 19 research articles; out of these, four articles in SCI-indexed journal and 18 articles in Scopus indexed journal. He is a reviewer of renowned journals such as IEEE Access, IEEE Consumer Electronics Magazine, IET Image Processing, IET Biometrics, Soft Computing, Imaging Science Journal, Signal Processing: Image Communication, and Computers and Electrical Engineering. His current research interests include Image Processing, Multimedia Security, Digital Watermarking, Artificial Intelligence, Medical Image Analysis, Biometrics and Compressive Sensing.

Chapter 1
Introduction to WiMAX System

The accomplished development in the utilization of advanced systems has prompted the requirement for the plan of new correspondence systems with higher capacity and data rates. The main inspiration has been the requirement of more bandwidth with lower latency in the operations. While throughput is the actual rate of data transfer, latency depends mostly on the processing speed of each node data streams go over through. While developing the new mobile technology, along with throughput-related performance enhancements, some similar parameters, like jitter, connectivity, scalability, interchannel interference, energy-efficiency, and compatibility with heritage networks, are also taken into consideration. The gigantic take-up rate of cell phone advancements, WLANs, and the exponential development that is encountering the utilization of the web has brought about an expanded interest for new techniques to acquire high limit remote systems. If too little detail is included in the model, one takes the risk of missing relevant exchanges and the resultant model does not provide up to the mark sympathetic.

1.1 Basic of System Modeling

This part portrays the exceptionally fundamental data of wireless system modeling which has been utilized for real data applications. A model is a disentangled show of the genuine framework planned for comprehension. Regardless of whether a model is a decent model or not relies upon the dimension to which it encourages the comprehension of the fundamental idea. Since all models are rearrangements of the real world, there is dependably an exchange off about what dimension of stuff is incorporated into the model. Simulation model should be reasonable fit for the

The original version of this chapter was revised. The correction to this chapter is available at
https://doi.org/10.1007/978-3-030-22460-8_7

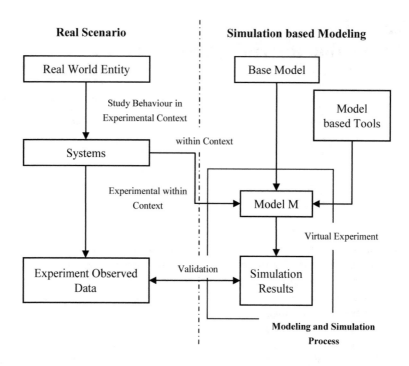

Fig. 1.1 Basic structure of system modeling

understanding of the concepts which designer wants to demonstrate. Model should have sufficient number of concept blocks as well it should be self explanatory to justify flow of the process.

Figure 1.1 demonstrates the proportionality between the real-time system and its prototype model displaying. Here it can be seen that by building up the model of a particular system, the virtual reality be accomplished. Fundamentally, the model is comparable to a virtual framework that is considered as a genuine element. For a continuous framework, the conduct can be determined by performing constant tests over it and its result can be seen as its reaction to specific information. Presently for any rising innovation which has been made sense of still on paper, the ongoing experimentation is absurd in the underlying phase of structuring. For that issue, the displaying would be the most appropriate model to investigate its conduct to a greatest degree. By developing the model of a system and performing simulation of it is more or less analogous to the experimentation performed in a real time system. In the event that the model satisfies every single imaginable detail, if the determined reproduction results surmised towards the test consequences of a framework then the model is said to have unsurpassed legitimacy and its yield results are viewed as practically same to that of the constant framework.

A simulation is the utilization of a model in such a means that it operates on time or space to compress it, thus facilitating one to make out the interactions that would not otherwise be clear because of their separation in time or space. It typically refers to a computerized translation of the model that is run over time to grasp the implications of the outlined interactions. Simulations are typically unvarying in their style. Here model is prepared and simulated to showcase given concept. Simulated model has been prepared conceptual block wise. Each block performs iterations as per requirement and pass over data in proper matrix to next block and all together stipulate output is generated. Simulation model gives actual idea of the real background and in that scenario how system may affect output that can be predicted virtually in the model.

Modeling and simulation are a method for developing grade of understanding of the interface of the elements of a system as well input output relationship shows performance of the system as a whole. The dimension of understanding which might be created by means of this procedure is only from time to time feasible through some other procedure. It is the utilization of models, including emulators, prototypes, simulators, and stimulators to develop data as a basis for making supervisory or technical decisions. The utilization of modeling and simulation inside engineering is all around perceived. Simulation technology belongs to the toolset of engineers of all application domains and has been incorporated into the body of data of engineering management. Modeling and simulation give complete idea regarding effect of present environment through which signal is going to pass. Moreover, it also gives detail idea of utilized frame work for all testing without implementing it in the hardware form [1].

1.2 WiMAX System and Various Design Technique of It

WiMAX might be viewed as the most developing remote systems administration standard endorsed by IEEE for understanding the improved portable framework situation as the union of cell communication, registering, Internet get to, and possibly numerous media applications becomes a genuine certainty. WiMAX is an IEEE 802.16 standard-based innovation in charge of carrying the broadband wireless access to the world as an option in contrast to wired broadband. It gives a proper answer for certain provincial access zones that are today kept from approaching wide band web as a result of cost thought.

Figure 1.2 demonstrates the WiMAX standard as an intermingling of two current wireless and cellular systems. At the end of the day, WiMAX fills the hole between the two rising patterns of things to come age. In a fixed wireless communication, WiMAX can replace the telephone company's copper wire networks, the cable TV's coaxial cable infrastructure while offering Internet Service Provider services. The WiMAX standard 802.16e gives fixed, traveling, convenient and versatile wireless broadband availability without the requirement for direct viewable pathway with the base station. Thus, WiMAX system which is becoming the perfect solution to

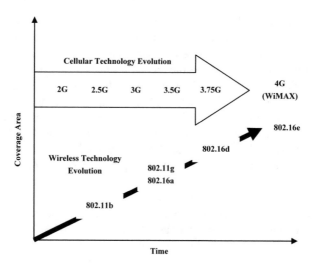

Fig. 1.2 Evolution of cellular and wireless technologies

meet the current demands of the future wireless networks thereby provides the tough competition to the existing 3G standards [2]. The diversity techniques and space-time coding (STC) are utilized to display the WiMAX framework.

Diversity is a ground-breaking correspondence transmitter-recipient system that gives wireless link improvement at a generally minimal effort. It exploits the irregular nature of radio spread by discovering independent signal path for correspondence. While managing the multi-way condition, the individual signal landing at the receiver faces free or profoundly uncorrelated fading. Today's reality, the principle objective of any wireless technology is to achieve highest system capacity with lower bit error rate which is unimaginable with a solitary transmitter and single getting reception apparatus since it cannot beat the impacts of fading. On the off chance that one radio way experiences a profound fading, another path may have a solid signal.

This development of independent fading in varied ways is exploited as a bonus to attain improved performance in wireless communication only if out of multiple paths; a minimum of one path is obtained with minimum distortion and maximum signal strength. This development leads towards the conception of diversity which can noticeably improve the performance over fading channels.

In a channel with multiple transmit or receive antennas spaced sufficiently so much enough; diversity may be obtained over area referred to as antenna diversity.

Two sorts of space diversity can be acquired to improve the limit of the framework. Transmit diversity in which multiple transmit antennae are used for the signal transmission which in term results in Multiple Input Single Output (MISO) diversity ($n \times 1$ system). While receive Diversity in which multiple receive antennae are used for the signal reception which in term results in Single Input Multiple Output

($1 \times n$ system). By utilizing various reception apparatuses at the two sides, the advantages of both the previously mentioned strategies can be experienced. This strategy is known as MIMO innovation.

In MIMO, by spatially allotting increasingly a greater number of antennas with adequate separation, the most extreme throughput can be acquired with least error rate. On the off chance that indistinguishable information will be transmitted through every one of the antennas, then at least one robust path may be recognized through which data would be received with the lowest errors and that is how the diversity gain can be achieved. Through this implementation of MIMO, the BER can be improved. To improve spatial limit regarding improved multiplexing gain, multiple antennas will have to pass various no. of data through spatial multiplexing process which is another encouraging element of the MIMO method.

Space-Time Coding (STC) is associate degree economical approach to use the diversity offered by the Multiple Input Single Output and Multiple Input Multiple Output. It is used to obtain gains due to spatial diversity via multiple transmit and receive antennas. Moreover, diversity gains proportional to the quantity of antennas at each transmit and receive side are often achieved. One well-liked illustration of these codes is the Alamouti scheme for two transmit antennas.

It has been as of now examined that the MIMO framework improves the performance of the system definitely by taking the benefit of antenna diversity techniques in the wireless communication system with a multipath situation. By transmitting the signal through numerous transmitter antennas which are set at a practically identical separation, the same signal can go in an alternate way and reach towards the receiver with the autonomous measure of fading relying upon the channel conditions that it has experienced. So out of two, one way may have offered less fading to information and it can be received with comparatively better accuracy which is what the unique feature of transmitter diversity. In any case, the upside of this component can be increased except if and until the antennas are put adequately far separated so that there would be no interference among the same information transmitted by multiple antennas. Presently, this factor puts the constraint on the limit of the framework. Because of some base separation necessity between two successive antennas, no. of antennas cannot be expanded beyond some limit and thus the diversity gain would be limited to that degree as it were. Here what the idea of Alamouti coding comes into the image.

In the execution of Alamouti coding with transmitter diversity, before transmission of same data through different multiple antennas, they get coded by taking their complex conjugates and afterward get transmitted. So now in the event that the partition between antennas is getting brought down, at that point additionally data cannot be interfered as a result of the coding and this is the manner by which diversity gain can be increased thereby anticipating much-improved execution. In this way, Alamouti coding technique with transmit diversity provides massive time and space diversity to the wireless communication system.

As the Orthogonal Frequency Division Multiplexing (OFDM) is the essential technology utilized in the WiMAX frameworks in the physical layer, it is basic to understand the nuts and bolts of this innovation and how it empowers high data rates to be supported in a wireless domain with NLOS task. All things considered, transmission conditions, multipath spread, and echoes from objects lead to arrive at the receiver in a time delayed fashion. These signals endure frequency selective fading because of the multi-path propagation impacts. At the point when a carrier is utilized to convey high information rates, which introduces enough delay and causes inter-symbol interferences by spreading symbols. Now in the situation of a single carrier modulation, this sort of engendering limits the information rates that can be utilized in Non-Line-Of-Sight (NLOS) situations. The technique of OFDM depends on the utilization of a large number of carrier's symbols spread in predefined bandwidth, with every carrier being modulated by a proportionately lower information rate than would be the situation with the single carrier transmission scenario. OFDM system hypothetically saves the bandwidth about 50% and gives a vigorous transmission strategy for NLOS situations.

1.3   Limitations of Existing Wireless System

There are two key wonders of wireless communication system that makes the issue challenging and intriguing. First is the phenomenon of fading: the varieties in the phase as well as time variation of the channel strengths due to the small-scale effect of multipath fading, as well as larger scale effects such as path loss via distance attenuation, shadowing, refraction or reflections by obstacles. Second, not at all like in the wired communication where every transmitter-recipient pair can frequently be recognized as an isolated point-to-point interface, wireless clients impart over the air range and there is interference between them in wireless communication. The interference can be between transmitters speaking with the single receiver, between sign from a single transmitter to multiple receivers or between various transmitter-receiver sets [3].

The WiMAX systems structure a significant piece of the wireless rollout of the future generation systems. They additionally give a substitution to major wired extensions of broadcast services, broadcast content feeder networks, and news-gathering networks accessible today by improving them with the new broadband highlights. Thusly, the WiMAX might be viewed as the last mile solution furnishing extremely high information rate alongside huge inclusion zone.

These days, different wireless administrators are currently looking for WiMAX innovation as the filling span between the current cellular framework and the future interest of high-speed communication with lower bit error rate. In general, WiMAX is arranged with the customary method for single transmitter and receiver antenna; however, it cannot actually frame the state of 4G technology. Now is the point at which the capability of WiMAX to grow an entirely new age of utilizations is at its prime. As talked about in the starting situation of the WiMAX framework, the most extreme research work is done in Single Input Single Output WiMAX system physical layer model and maximum data throughput received accordingly. However, in the present situation, during the period of real-time voice or image transmission through WiMAX framework, the accessible bit error rate and signal to noise ratio and consequently the capacity of system are serious limitations for the implementation. So, in the 4G transmission system, link reliability and maximum data throughput are the need for transmitting the real-time voice as well as the image information at high speed. Implementation of various antenna diversity techniques along with OFDM technique is one of the promising solutions for this. So, it is very important to design WiMAX system available for real-time data transmission (such as image and speech) to achieve the lower bit error rates, higher signal to noise ratio, and higher system capacity.

1.4 Motivation and Important Points Covered in the Book

These days, life does not appear to be doable without wireless systems in either structure. Wireless technologies will not be limited only for the mobile communication system but it will cover almost all the sectors such as various industries, semiconductor vendors, various manufacturing firms and many more. Next generation, technologies will reach not only for cell phones but also it will reach to almost all industries which are connected with wireless transmission. Huge demand for wireless based services such as to carry video and other rich content services and IoT (internet of things) based services are the major push towards the race to next generation technology. To fulfill these major objectives, next generation technology will provide large broadband speed, ultrareliable connectivity, and ultralow latency for minimum delay in the communication. Wireless is becoming the leader in com-

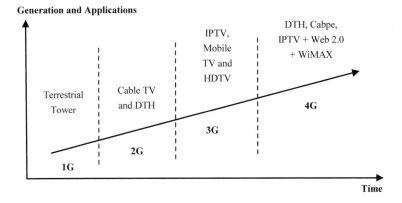

Fig. 1.3 Scenarios of wireless system generation

munication choices among users as justified from Fig. 1.3 which depicts the growth of wireless communication on the scale of time in terms of generations. It is not anymore, a backup solution for nomadic travelers but really a mind-set normally used everywhere even when the wired communications are possible. Wireless access has been accessible to us for a long time now. It's most obvious indication that it is in the form of remote LANs and the Wi-Fi hotspots. Usage of Wi-Fi and wireless LANS are with PCs, PDAs and different gadgets in the various premises in certain distance.

The clients can peruse the web; make VoIP calls utilizing programming, for example, Skype, get to mail, or transfer pictures and recordings from computerized cameras. They can likewise watch the video by gushing from any of the video sources or downloading video files. The guarantee for broadband portable administrations keeps on developing. Normally, rapid broadband arrangements depend on wired-get to advances, for example, computerized endorser line (DSL). This kind of arrangement is not anything but difficult to send in remote rustic regions, and besides, it needs support for terminal versatility. Likewise, the continuous advancement in the utilization of remote systems has prompted the prerequisite for the structure of new present-day correspondence systems with higher limit and lower mistake rate. The media transmission industry is likewise updating, with a necessity for a more prominent scope of administrations, for example, video gatherings, or applications with sight and sound substance. The expanded reliance on PC organizing and the internetwork has brought about a bigger interest for associations with be distributed whenever, prompting an expansion in the prerequisites for more noteworthy limit and extremely dependable broadband wireless communication systems.

Presently in the cutting edge for wireless communication, portable terminals will face mass information administration. Along these lines, the signal procedure in the portable terminal is expected to economize power, while high spectrum efficiency and network reliability should be ensured in the meantime. For this issue, new advancements with high throughput with less requirement on bandwidth have been planned. In actuality, the prerequisites on bandwidth and spectrum availability are unending.

Thus, the different firms associated with the area of wireless communication faces troubles to satisfy the necessity of transfer speed for effective and exact transmission and reception. Additionally, the issues of time-varying nature of channels, for example, fading and multipath, put the confinement on the performance of high information rate with good quality of service. The demands for greater capacity, high reliability as well as accuracy are the prime imperatives for the anticipated ages of the wireless systems, for example, Wi-Fi, WiMAX, and so on.

The book displayed a presentation examination of the latest wireless system innovations, for example, WiMAX alongside its physical layer working which depends on the Orthogonal Frequency Division Multiplexing (OFDM) technique. In spite of being an almost 50-year-old idea, it is just in the most recent decade that OFDM turns into the decision of the modem in wireless applications. One of the biggest advantages of an OFDM is the capacity to change over dispersive broadband channels into parallel narrowband sub-channels, in this manner essentially simplifying equalization at the receiving end. The fundamental OFDM method alongside MIMO implementation is utilized to build the system capacity by lessening the impacts of ISI.

First of all, the simulation and performance analysis of OFDM technique is presented for the physical layer functioning of WiMAX system in this book. Then after the modeling of physical layer, single antenna WiMAX system is presented and simulation results are displayed mainly in terms of input data, coding techniques as per IEEE 802.16 standards, digital modulation techniques, OFDM technique, and channel SNR and bit error rate.

After modeling of WiMAX system along the physical layer for the OFDM technique, the simulation and performance analysis of various diversity techniques for WiMAX system is presented in this book. Transmit diversity is one the best scheme to improve system capacity and reduce BER. The space time coding, i.e. Alamouti coding can be utilized to implement that and for performance improvement. Also, critical comparative analysis of traditional single antenna system and multiple antenna systems for real-time modeling of WiMAX system are presented in this book. Finally, to enhance the BER and the capacity of the WiMAX framing, the different transmitter and receiver diversity techniques are presented in WiMAX model with the Alamouti coding method. The software implementation of the WiMAX model for real-time transmission of image and speech signals is also presented in this book. The important points covered in this book are listed below:

- Provide information about various designing techniques for WiMAX system modeling as per 4G standard requirements.
- Provide a complete WiMAX system model according to the IEEE 802.16 standards.
- Provide analysis of different features of antenna diversity techniques in wireless communication to overcome limitations due to multipath fading.
- Provide a simulated WiMAX system using transmitter and receiver diversity techniques along with Alamouti coding scheme for the improvement of bit error rate, thereby rising system reliability.
- Provide virtual reality of WiMAX system by modeling the IEEE 802.16 standard for transmission of real-time data such as digital images and speech signals.

1.5 Organization of Book

The present section talks about the fundamental comprehension of modeling and simulation of system alongside a review of WiMAX framework using antenna diversity techniques and Alamouti coding method. The remaining chapters talk about the modeling and simulation of WiMAX framework dependent on IEEE

802.16 standard and software implementation of various antenna diversity techniques in it.

Chapter 2 incorporates the total hypothetical investigation of the WiMAX framework along with Wi-Fi and cellular system. The distinctive WiMAX guidelines alongside the highlights have been talked about. At the last phase of this section, the conventional model of WiMAX framework has been actualized by the IEEE 802.16-2004 standard as a centerpiece of this part in which the impact of time diversity can be seen by simulating the model with and without the effect of FEC coder. Besides, the impact of frequency diversity can be examined by changing the cyclic prefix of OFDM sub-block of the WiMAX framework. As the most significant comment, it very well may be found from the modeling of traditional WiMAX framework that the performance of WiMAX system in terms of BER is absolutely reliant on channel SNR and it is unrealistic to build the SNR beyond certain breaking point to improve the performance of system. This impediment defeats by planning model exhibited in Sect. 1.3.

Section 1.3 presents execution of different antenna diversity techniques and OFDM for WiMAX framework model. The principle confinement of WiMAX found from second section that its performance is limited to channel SNR range. This impediment has been overwhelmed with the implementation of space diversity. The utilization of multiple antennas at the transmitter as well as at the recipient in a communication connection opens another measurement in reliable wireless communications, improving the performance of the framework generously. This section incorporates the MATLAB algorithms and simulations of various antenna diversity strategies. Based on the number of transmitting and receiving antennas used in the wireless system Alamouti coding can be modified and implemented in effective way. Additionally, the impact of changing modulation order over system BER has been observed. Finally, the analysis of OFDM, which is the base of physical layer, has been carried out.

Section 1.4 manages the designing and simulation of quality-based WiMAX framework alongside the impact of time, frequency, and space diversity strategies. In this part, the SNR bound impediment of traditional WiMAX framework has been overwhelmed by the execution of antenna diversity algorithms. In general, this section assesses the different impacts of various diversity techniques and their parameters over BER of WiMAX framework so as to obtain the optimum values of the same for the real-time implementation of WiMAX system.

Chapter 5 includes the entire WiMAX framework by-passing real-time image and speech signals along with the implementation of various diversity techniques like time, frequency, and space. The creative touch to the total model has been given by-passing real-time image and speech signals through it. In the following back to back stages, the impact of antenna diversity techniques with/without Alamouti coding over system BER has been observed with again real-time image and speech transmission. All in all, towards the finish of this section, the total WiMAX framework demonstration can be gotten with the most efficient methodology with least error rate. At last, the concluding summary of book and future research heading are discussed in Chap. 6.

References

1. Petrone, G., & Cammarata, G. (2008). *Recent advances in modelling and simulation*. Vienna: I-Tech Education and Publishing.
2. Yarali, A., Mbula, B., & Tumula, A. (2007, March). WiMAX: A key to bridging the digital divide. In *Proceedings 2007 IEEE SoutheastCon* (pp. 159–164). IEEE.
3. Tse, D., & Viswanath, P. (2005). *Fundamentals of wireless communication*. Cambridge: Cambridge university press.

Chapter 2
WiMAX Introduction and Modeling

2.1 Introduction

WiMAX (Worldwide Interoperability for Microwave Access) is the very projected innovation that means to give business and consumer wireless administrations in type of Metropolitan Area Network (MAN). The mobile WiMAX is a brilliant creation which is satisfying the most recent interest. Through its high coverage and information rate qualities, it satisfies total system design in this manner giving an adaptable and cheap solution for the last mile. The interoperability is an extremely critical issue, on which gear cost and volume of sales will be based. Administrators won't be bound to a sole hardware provider, as the radio base stations will probably interact with terminals created by various providers. From the perspective of expense and exactness, the clients must get the advantage of the provider's competition. WiMAX might be viewed as advanced fourth generation (4G) of mobile system as the convergence of cellular communication, Internet access, and possibly numerous interactive media applications becomes a genuine certainty [1].

WiMAX's ascribes open the innovation to a wide variety of utilizations. With its enormous range and high transmission rate, WiMAX can fill in as a spine for 802.11 hotspots for connecting to the Internet. Then again, clients can interface mobile devices, for example, PCs and handsets straightforwardly to WiMAX base stations without utilizing 802.11 which can be all around seen from Fig. 2.1. Designers venture this setup for the WiMAX portable form, which will give clients broadband availability over large coverage areas compared with 802.11 hotspots' moderate coverage. Mobile devices associated legitimately to WiMAX base stations likely will accomplish a range of 5–6 miles and the technology can also provide fast and cheap broadband access to markets that lack infrastructures (fiber optics or copper wire), such as rural areas and unwired countries.

Currently, several companies offer proprietary solutions for wireless broadband access, many of which are expensive because they use chipsets from adjacent technologies, such as 802.11. Manufacturers of these solutions use the physical layer

© Springer Nature Switzerland AG 2020
B. S. Sedani et al., *WiMAX Modeling: Techniques and Applications*,
https://doi.org/10.1007/978-3-030-22460-8_2

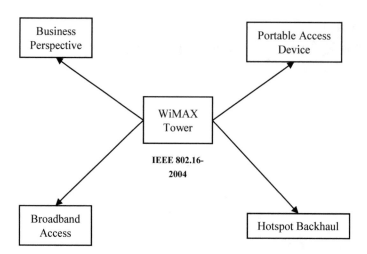

Fig. 2.1 WiMAX scenario

and bypass the medium access control layer by designing a new one. Unlike these proprietary solutions, WiMAX's standardized approach offers economies of scale to vendors of wireless broadband products, significantly reducing costs and making the technology more accessible. Many companies that were offering proprietary solutions, however, have participated in the WiMAX forum and now offer WiMAX-based solutions. WiMAX can be used in disaster recovery scenes where the wired networks have broken down. Similarly, WiMAX can be used as backup links for broken wired links. Additionally, WiMAX will represent a serious competitor to 3G cellular systems as high-speed mobile data applications will be achieved with the 802.16e specification [2]. The main operators have concentrated their interests and efforts on the future applications of this new technology. The WiMAX forum created in April 2002 is a no-profit organization that groups companies promoting the broadband access based on the wireless communication standard, point to multi-point IEEE 802.16 for Metropolitan Area Network. WiMAX forum activities aim to:

- Support the standardization process of IEEE 802.16 for MAN
- Select and promote some of the WiMAX profiles defined in the 802.16
- Certificate the interoperability between WiMAX equipment of different suppliers
- Make WiMAX a universally accepted technology

Several corporations that were providing proprietary solutions, however, have participated within the WiMAX forum and currently supply WiMAX based mostly solutions. WiMAX are often employed in disaster recovery scenes wherever the wired networks have counteracted. Similarly, WiMAX are often used as backup links for broken wired links. The WiMAX forum created in Apr 2002, could be a no-profit organization that teams corporations promoting the broadband access supported the wireless communication normal, purpose to multipoint IEEE 802.16 for Metropolitan space Network. WiMAX forum activities aim to:

- Hold up the standardization of IEEE 802.16 for MAN
- Support a number of the WiMAX framework outlined within the 802.16
- Design WiMAX technology in such a way that it can be accepted universally [2]

2.2 Relationship with Other Wireless Technology

Wireless access to information systems is relied upon to be a territory of quick development for versatile communication frameworks. The immense take-up rate of cell phone advances, WLANs, and the exponential development that is encountering the utilization of the web has brought about an expanded interest for new techniques to get high capacity wireless systems. WiMAX is expected to have an explosive growth, as well as the Wi-Fi, but compared with the Wi-Fi, WiMAX provides broadband connections in greater areas, measured in square kilometers, even with links not in LOS. Hence WiMAX is a MAN, featuring that "metropolitan" is alluded to the expansion of the territories and not to the density of populace. Be that as it may, Wi-Fi and WiMAX are not competing for technologies. While WiMAX can give high limit web access to homes and business seats, Wi-Fi permits the expansion of such associations inside the corporate site's buildings [3].

Figure 2.2 set out the relative stage among three present-day wireless innovations for example, WiMAX, Wi-Fi, and 3G cell communication. Regardless, both WLAN and cellular applications are extended to offer the requested wireless access. In any case, they experience a few troubles for achieving total versatile broadband access, limited by factors, for example, transmission capacity, coverage area, and foundation costs.

As appeared following Fig. 2.3, Wi-Fi gives a high data rate, however just on a short scope of separations and with a slow movement of client. Then again, cellular system offers high ranges and vehicular versatility; however rather, it gives lower data rates and requires high speculations for deployment. WiMAX attempts to adjust this circumstance which is pictorially portrayed in Fig. 2.3. WiMAX fills the gap between Wi-Fi and cellular, hence giving vehicular mobility, high service areas, and high information rates.

2.3 WiMAX Standards

WiMAX is an innovation institutionalized by IEEE for wireless MANs conforming to parameters which empower interoperability. WiMAX improvements have been moving forward at a quick pace since the underlying institutionalization endeavors in IEEE 802.16. Meanwhile, the metropolitan area wireless systems improvement

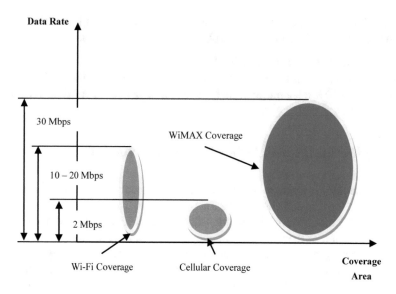

Fig. 2.2 Relationship of WiMAX with other wireless technologies

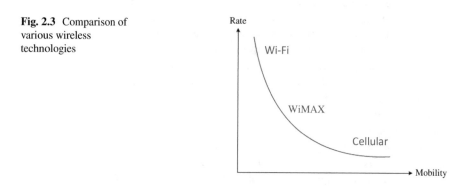

Fig. 2.3 Comparison of
various wireless
technologies

work was advancing under the IEEE 802.16 advisory group which developed
benchmarks for wireless MANs. The IEEE 802.16 standard was firstly specified to
address correspondences with direct visibility in the frequency band from 10 to
66 GHz. Because of the way that NLOS transmissions are troublesome when
imparting at high frequencies, the correction 802.16a was determined for working
in a lower frequency band, somewhere in the range of 2 and 11 GHz [4].

The IEEE 802.16d determination is a variety of the fixed standard (IEEE 802.16a)
with main benefit of improving the power utilization of the cell phones. Standards
for fixed WiMAX (IEEE 802.16-2004) were announced in 2004, trailed by mobile
WiMAX (IEEE 802.16e) in 2005. Then again, the IEEE 802.16e standard is a correction to the 802.16-2004 base determination with the aim of targeting the mobile
market by adding portability. WiMAX standard-based items are intended to work
with IEEE 802.16-2004 as well as with the IEEE 802.16e in particular. While the

Table 2.1 WiMAX standards

WMAN standard	Definition year	Frequency band (GHz)
IEEE 802.16	2001	10–66
IEEE 802.16(a)	2003	2–11
IEEE 802.16(b)	2003	5–6
IEEE 802.16(c)	2003	10–66
IEEE 802.16(d)	2003	2–11
IEEE 802.16-2004	2004	2–11
IEEE 802.16(e)	2005	2.3–3.4

802.16-2004 is fundamentally proposed for stationary transmission, the 802.16e is arranged to both stationary and mobile organizations. The WiMAX forum, an industry body established in 2001 to elevate conformance to models and interoperability among remote MAN systems, at that point delivered the WiMAX as today is usually known. In Europe, the gauges for wireless MANs were formalized as Hiper MANs. These were likewise founded on IEEE 802.16 principles yet did not at first utilize similar parameters, (for example, frequency or number of subcarriers). These were later blended with the WiMAX models. The IEEE 802.16d measures accommodate fixed and traveling access, while the 802.16e principles likewise give versatility up to velocities of 120 km for every hour. The short outline of WiMAX benchmarks is given in Table 2.1.

2.4   Technical Overview of WiMAX Standards

In this section, technical information on WiMAX standards is discussed. The WiMAX standard specifies the air interface for the IEEE 802.16-2004 specification working in the frequency band 2–11 GHz. This air interface describes the definition of the medium access control (MAC) and the physical (PHY) layer.

2.4.1   Medium Access Control (MAC) Layer

A few functions are devoted to giving administration to supporters that incorporate transmitting information in edges and controlling access to the common wireless medium. The medium access control (MAC) layer, which is arranged over the physical layer, bunches the referenced functions. The first MAC is upgraded to make various physical layer specifications and services, tending to the requirements for various situations. It is commonly intended to work with point-to-multipoint topology systems, with a base station controlling independent sectors all the while. Access and bandwidth allotment calculations must almost certainly require many terminals per channel, with terminals that might be shared by various end clients. In this manner, the MAC convention characterizes how and when a base station (BS)

Table 2.2 MAC layer header fields

Type	Length (bits)	Function
CI	1	CRC bit indicator
		1 = CRC incorporated in the PDU by appending it to the PDU payload after encryption, if any
		0 = no CRC is included
CID	16	Connection identifier
EC	1	Encryption control
		1 = means encrypted payload
		0 = payload is not encrypted
EKS	2	Encryption Key Sequence
HCS	8	Header Check Sequence
		Detection of errors
HT	1	Header type
		Shall be set to zero
LEN	11	The length in bytes of the MAC PDU as well as the MAC header and the CRC if present
Type	6	This field describes the subheaders and special payload types

or subscribers' station (SS) may start the transmission on the channel. To accomplish synchronization during the transmission-gathering process, an aggregate of 48 overhead bits condensed in Table 2.2 are included alongside the data frame as a preamble.

The "CI" bit specifies the presence of a CRC code for the error checking task. "CID" forms the 16-bit data for determining the connection. "EC" bit justifies whether the data is encrypted or not. "HCS" and "HT" specify the characteristic of the header field. "LEN" shows the length of whole MAC PDU. The last field "Type" describes the subheader.

2.4.2 Physical Layer

The IEEE 802.16-2004 standard characterizes three distinctive PHYs that can be utilized related to the MAC layer to give a reliable end-to-end link. This PHY layer characterizes the accompanying determinations:

1. **Randomizer**: Randomization is the primary procedure done in the physical layer after the data packet is gotten from the MAC layer. Randomizer drives on a bit by bit premise. Each burst in transmitter and receiver is randomized. The reason for the scrambled data is to change over long arrangements of 0's or 1's in its irregular grouping to improve the coding performance. The primary segment of the data randomization is a pseudo random binary sequence generator. This generator is implemented by utilizing a linear feedback shift register.

2. **Time Diversity with Forwarded Error Correction**: Diversity in time is given through forward error correction which is done in transmitter and receiver and comprises of the connection of Reed–Solomon external code and a rate compatible convolutional inner code. The reason for utilizing a Reed–Solomon code to the information is to add the redundancy to the data sequence. This redundancy helps in correcting block errors that happen during transmission of the signal. In WiMAX physical layer, the Reed–Solomon outer code is encoded by the inner convolutional encoder. Convolutional codes are utilized to correct random errors in information transmission.

3. **Block Interleaving**: Interleaving in its most fundamental structure can be depicted as randomizer yet it is very unique in relation to the randomizer as it does not change the condition of the bits; however, it basically works on the position of bits. Interleaving is finished by spreading the coded symbols in time before the modulation process in transmitter, and de-interleaving is completed at receiver side after the demodulation process.

4. **M-QAM Technique**: The interleaver reorders the information and sends the information frame to the M (Modulo)-QAM block. The function of the M-QAM is to map the approaching bits of information from interleaver onto a group of signals, i.e., constellation. In the transmitter stage, the coded bits are mapped to the IQ constellation and data bursts are transmitted with equivalent power by utilizing a normalization factor.

5. **Frequency Diversity with OFDM Technique**: Frequency diversity is master vided by OFDM method which permits the transmission of multiple signals using different subcarriers simultaneously. Since the OFDM waveform is made out of numerous narrowband orthogonal carriers, selective fading is localized to a subset of carriers that are relatively easy to equalize.

6. **Space Diversity in Fading Environments**: Optional help of both transmitter and receiver to enhance the performance in fading situations through spatial diversity, enabling system to increase capacity. The transmitter executes space-time coding (STC) to give transmit source independence, decreasing the fading margin necessity, and battling interference. The receiver, notwithstanding, utilizes Maximum Ratio Combining (MRC) techniques to improve the accessibility of the system.

2.5   Important Features of WiMAX Standards

The important features of WiMAX standards are pointed out as per below:

1. **OFDM Based Physical Layer**: The WiMAX physical layer depends on OFDM, which is an exquisite and viable strategy for defeating multipath distortion.

2. **Very High Peak Data Rates**: WiMAX is equipped for supporting high information rates. Truth be told, the peak PHY information rate can be as high as 70 Mbps when working utilizing a 20 MHz wide spectrum.

3. **Orthogonal Frequency Division Multiple Access (OFDMA)**: Mobile WiMAX utilizes OFDM as a various access strategy, whereby various user in the network can be allotted with various subsets of the OFDM tones. OFDMA encourages the exploitation of frequency diversity and multi-user diversity to fundamentally improve the framework capacity.

4. **Adaptive Modulation and Coding (AMC)**: WiMAX bolsters various propelled signal-preparing procedures to improve overall system capacity. These techniques incorporate adaptive modulation and coding, spatial multiplexing, and multi-user diversity.

5. **Link Layer Re-transmissions**: For data connections that require enhanced reliability, WiMAX bolsters automatic re-transmission request (ARQ) at the link layer. ARQ empowered connections require each transmitted parcel to be acknowledged by the receiver; unacknowledged bundles are thought to be lost and are re-transmitted.

6. **Support for Advanced Antenna Techniques**: The WiMAX arrangement has various guides incorporated with the physical layer plan, which takes into consideration the utilization of numerous radio antenna techniques. WiMAX offers high spectral proficiency, especially when utilizing higher order MIMO arrangements.

7. **Quality-of-Service Support**: The WiMAX MAC layer has a connection situated design. WiMAX has a truly adaptable MAC layer that can accommodate a variety of traffic types, including image, speech, video, and multimedia by providing strong quality of service.

8. **Robust Security**: Robust security capacities, for example, solid encryption and mutual authentication, are incorporated with the WiMAX standard.

9. **Internet Protocol (IP) Based Architecture**: WiMAX characterizes an adaptable all-IP based architecture that allows for the exploitation of all the advantages of IP. The reference system model requires the utilization of IP-based protocols to convey end to end capacities, for example, QoS, security, and mobility management.

As yet, every part of WiMAX structure, layering approach, functions of every individual subsystem just as their characteristics have been very much talked about. Presently dependent on this talk, the forthcoming sections described the real demonstration and experimentation of the WiMAX framework for seeming well and good. In segment 2.6, the fundamental modeling of the physical layer for traditional WiMAX framework has been examined while in Sect. 2.7, the image has made to go through the WiMAX model envisioning that the future age WiMAX framework is likewise skilled to pass picture alongside just information.

2.6   Traditional WiMAX System Modeling

In this part, traditional WiMAX framework displaying is examined by utilizing Simulink tool of MATLAB. Exhibited model in this section is a straightforward traditional WiMAX framework and after that it will be extended towards implementation of advanced antenna system by reproducing different antenna diversity strategies in essential WiMAX framework. Figure 2.4 demonstrates a basic structure of WiMAX framework which has been broadly utilized. The model itself comprises of three primary segments to be specific transmitter, receiver, and channel. Transmitter and receiver segments comprise of source encoder and decoder, channel encoder and decoder, modulator and demodulator sort of sub-frameworks while channel is demonstrated as Additive White Gaussian Noise (AWGN) channel. The accompanying subsections delineate the block by block description of every subsystem for conventional WiMAX framework with a single transmitter and single receiver antenna.

Figure 2.5 illustrates the basic architecture of a traditional WiMAX system. The WiMAX framework includes transmitter, receiver, and wireless channel. Here the experimentation has been done by selecting random data from a randomizer. The performance analysis of the WiMAX framework has been evaluated in terms of BER v/s SNR.

2.6.1   WiMAX Transmitter

The WiMAX transmitter contains mostly two layers activity. One is the MAC layer for the different security algorithms and second is the physical layer which is the inside purpose of this book. Figure 2.6 demonstrates the snapshot of WiMAX transmitter framework which includes the input information unit as a piece of MAC layer and randomizer, series of coders, QAM as a coder, and OFDM unit as parts of the physical layer.

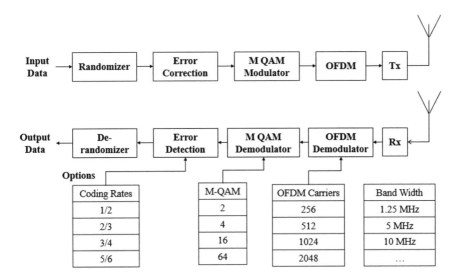

Fig. 2.4 Basic architecture of WiMAX system

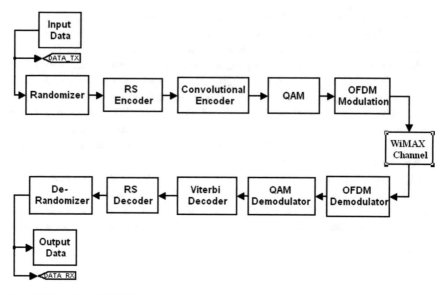

Fig. 2.5 Traditional WiMAX system model

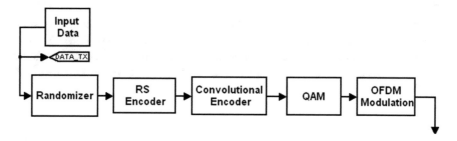

Fig. 2.6 WiMAX transmitter

2.6.1.1   Input Data

This unit essentially goes about as a data generator unit which is exceedingly required for the virtual transmission of information through the WiMAX system. As appeared in Fig. 2.7, the random integer block creates the double integer informa- tion data with the measurement (29×1) which would be changed over (232×1) bits by integer to bit converter that would be combined with 46 overhead bits plus two reserved bits possessed by the MAC layer of the WiMAX system. MAC layer man- ages the security algorithms. In Fig. 2.7, the header arrangement of MAC informa- tion is additionally demonstrated where every single field is described by their distinct function and length. Prior to transmission of data bits, the information would be appended with 48 bits as a piece of MAC header for the security purposes. At that point after the processing of bit stream of dimension (280×1) as a transmit

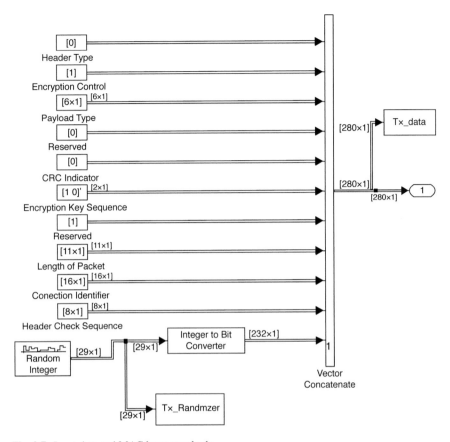

Fig. 2.7 Input data and MAC layer standards

information further stream would be prepared further. Moreover, it is indicated by the block of WiMAX transmitter.

2.6.1.2   Randomizer

The role of the source in the WiMAX framework is played by the unit known as randomizer. The data bits must be randomized before the process of real-time transmission. The randomization procedure is utilized to limit the likelihood of transmissions of non-modulated subcarriers. Figures 2.8 and 2.9 demonstrate the PN sequence instatement alongside the structure of actual randomizer in the WiMAX framework model, respectively. The procedure of randomization is performed on each burst of information on the downlink and uplink, and on every allocation of an information block.

For this situation, rather than performing out a randomization procedure, a binary source that produces arbitrary groupings, i.e., random sequence of bits, is utilized.

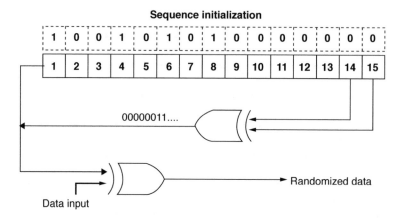

Fig. 2.8 Initialization of PN sequence

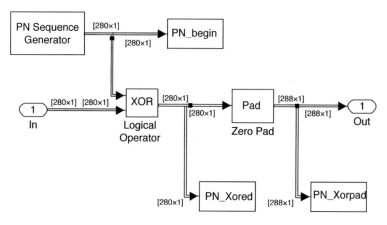

Fig. 2.9 PN sequence initialization in WiMAX modeling

The quantity of bits that are created is indicated to outline based and is determined from the packet size required in every circumstance. The bundle size relies upon the quantity of transmitted OFDM images and the general coding rate of the framework, just as the tweak letters in order. The flood of downlink parcels will be randomized by modulo-2 expansion of the information with the yield of pseudo random binary sequence generator. Here, the size of the information bit produced is 280 bits in the PRBS. Towards the start of each burst, the PRBS register is cleared and the seed estimation of 100101010000000 is stacked. Pseudo random binary sequence generator creates a profoundly arbitrary code and it is used for making the information excessively extremely random so just the similar sequence can decode the code; some other unauthentic client cannot get a handle on the data after the information has been converged with this code. The seed worth can be utilized to calculate the randomization bits, which are consolidated in an XOR activity with the serialized bit stream of each burst. These (280 × 1) bits are padded with 8 zero bits for the further procedure of encoding and modulation.

2.6.1.3   Reed–Solomon (RS) Encoder

The encoding procedure comprises of a connection of an outer Reed–Solomon (RS) code and an inner Convolutional Code (CC) as a FEC conspires. That implies that first information goes in square arrangement through the RS encoder, and after that, it goes over the convolutional encoder.

It is an adaptable coding process because of the puncturing of the signal and permits distinctive coding rates. The last part of the encoder is a procedure of interleaving to stay away from long error bursts. The properties of Reed–Solomon codes make them appropriate to applications where the errors occur in burst. Reed–Solomon error rectification is a coding plan which works by first building a polynomial from the information symbols to be transmitted, and after that sending an oversampled version of the polynomial rather than the original symbols themselves. A Reed–Solomon code is indicated as RS (n, k, t) with l-bit symbol. This implies that the encoder takes k information symbols of l bits each and adds $2t$ parity symbols to develop an n-symbol codeword. In this way, n, k, and t can be characterized as:

- n: number of bytes after encoding process
- k: number of data bytes before encoding process
- t: number of data bytes that can be corrected

The blunder adjustment capacity of any RS code is controlled by $(n - k)$, the proportion of redundancy. In the event that the location of the wrong symbol is not known ahead of time, at that point a Reed–Solomon code can correct up to t symbols, where t can be communicated as $t = (n - k)/2$. As indicated in the standard, the Reed–Solomon encoding will be derived from a systematic RS ($n = 255$, $k = 239$, $t = 8$) code utilizing a Galois field determined as GF (2^8). The primitive and generator polynomials utilized for the systematic code are communicated as pursues:

Primitive Polynomial: $P(x) = x^8 + x^4 + x^3 + x^2 + 1$
Generator Polynomial: $g(x) = (x + \lambda^0)(x + \lambda^1)(x + \lambda^2)...(x + \lambda^{2t-1})$

Now as shown in snapshot of the block shown in Fig. 2.10, initially the (288 × 1) bits have been converted into (36 × 1) integers because the RS encoder can route over the integer information. This information would be padded with zero bits to form data of (239 × 1) double values as the data length of RS encoder is taken to be $k = 239$.

According to the way towards coding clarified over the yield of (255 × 1) whole integers would be obtained. These data tests have been basically changed over to vector of (40 × 1) by the help of block "U-Y selector" on the grounds that after that (320 × 1) bits can be acquired through the block "integer to bit converter" as the number of bits per integer has been set to 8 as a particular parameter. These data bits

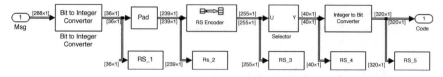

Fig. 2.10 Simulation model for RS encoder

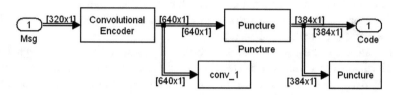

Fig. 2.11 Simulation model for convolution coder

have been additionally prepared to inner coding through the block of convolution encoder that has been clarified in the following subsection.

2.6.1.4 Convolution Encoder

After the completion of RS encoding procedure, the data bits are additional encoded by a binary convolutional encoder, which has a native rate of 5/6 and a constraint length of 7. Now in this case, the generator polynomials used to derive its two output code bits represented as X and Y are defined in the following expressions:

$$G1 = 17_{OCT} \quad \text{for X}$$
$$G2 = 133_{OCT} \quad \text{for Y}$$

A convolutional encoder receives information of length k_0 bits and produces codewords of n_0 bits. Generally, it is designed of a shift register of L segments, where L denotes the constraint length. Figure 2.11 represents the construction of the inner convolution code.

Here as the native rate of the coder has been chosen as 5/6, so by applying input (320×1), the output of (640×1) has been observed. These (640×1) bits are given to the block of "puncture vector" which gives output as (384×1). Puncturing is the process of systematically deleting bits from the output stream of a low rate encoder in order to reduce the amount of data to be transmitted, thus forming a high-rate code. The bits are deleted according to a perforation matrix, where a "zero" means a discarded bit. The process of puncturing is used to create the variable coding rates needed to provide various error protection levels to the users of the system. The different rates that can be used are rate 1/2, rate 2/3, rate 3/4, and rate 5/6. The puncturing vectors for these rates are given in Table 2.3. This punctured output has been modulated by quadrature amplitude modulation block.

Table 2.3 Puncturing vectors for different rates

Rate	Puncture vector
1/2	[1]
2/3	[1 1 1 0]
3/4	[1 1 0 1 1 0]
5/6	[1 1 0 1 1 0 0 1 1 0]

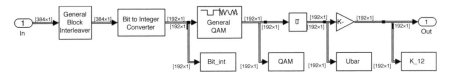

Fig. 2.12 Simulation model for Quadrature Amplitude Modulator (QAM)

2.6.1.5 Quadrature Amplitude Modulator (QAM)

When the sign has been coded, it enters the modulation block. All wireless communication frameworks utilize a modulation scheme to map coded bits to a structure that can be successfully transmitted over the wireless channel. In this manner, the bits are mapped to a subcarrier amplitude and phase, which is represented by a complex In-phase and Quadrature-phase (IQ) vector. The IQ plot for a modulation scheme demonstrates the transmitted vector for all information word blends. Gray coding is a technique for this designation so that adjacent points in the constellation only differ by a single bit. This coding technique helps to minimize the overall bit error rate as it reduces the chance of different bit errors producing from a single symbol error.

Figure 2.12 demonstrates the essential structure of quadrature amplitude modulation (QAM) technique which includes various no. of blocks that can process integer type of data only. Before changing over the approaching double bits (384×1) into the integer value (192×1) with the end goal of quadrature amplitude modulation, the bits are experienced through the way towards interleaving. This block of a framework can also be projected as a piece of the encoder sequence.

Information interleaving is commonly used to scatter error bursts and, in this manner, reduce the error focus to be remedied to expand the efficiency of FEC by spreading burst error presented by the transmission channel over a longer time. At the point when the block of errors is consistently experienced, around then the impacts of errors can be limited by the way towards interleaving where the data in form of the matrix will be manipulated by interchanging lines into segments, i.e., rows into column and the other way around. Interleaving is ordinarily executed by utilizing a two-dimensional array buffer to such an extent that the information enters the buffer in lines, which indicate the number of interleaving levels, and after that, it is read out in specific columns. The outcome is that a burst error in the channel subsequent to interleaving becomes few scarcely spaced single symbol errors, which are more easily correctable.

WiMAX utilizes an interleaver that consolidates information utilizing 12 interleaving dimensions. The impact of this procedure can be comprehended as a spreading of the bits of the various symbols, which are consolidated to get new symbols, with a similar size yet with rearranged bits. The interleaver of the test system has been actualized in two stages. In the first place, information goes through a matrix interleaver which performs block interleaving by filling a matrix with the information symbols row by row, and after that sending the content column by column. The parameters utilized for this block are number of rows and columns that make the matrix: $N_{rows} = 12$, total no. of coded bit, i.e., $N_{cbps} = 384$, and the quantity of coded bits per subcarrier for example $N_{cpc} = 2$ for 4-QAM modulation strategy according to the code of the exhibited framework.

$$N_{columns} = \frac{N_{tcb}}{N_{rows}} \tag{2.1}$$

The second step consists of a block interleaver. It rearranges the elements of its input according to an index vector. This vector is defined as:

$$I = \sum_{i=0}^{N_{tcb}-1} \left(s\frac{i}{s} + \mod\left(i + N_{tcb} - \frac{iN_{rows}}{N_{tcb}}, s \right) + 1 \right) \tag{2.2}$$

In this case, N_{tcb} describes the total number of coded bits, $N_{tcb} = N_{cpc} \times N_{tx}$ is a data, N_{cpc} describes the number of coded bits per subcarrier, and $s = N_{cpc}/2$.

MATLAB logic behind the implementation for this presented work is as per below:

```
Ncbps = 384; Ncpc = 2;
k = 0: Ncbps - 1
mk = (Ncbps/12) * mod(k,12) + floor(k/12);
s = ceil (Ncpc/2);
jk = s * floor(mk/s) + mod (s, mk + Ncbps - floor (12 * mk/Ncbps));.
```

After the process of block interleaving, the manipulated information regarding bits gets changed over into (192×1) integers with the end goal of quadrature modulation as it is the essential of QAM block. The "number of bits per integer" parameter of the block characterizes what numbers of bits are mapped for each output. In this case, this parameter equivalent to 2 has been chosen that is the reason the vector of (384×1) has been changed over to a vector of (192×1) which would be the output of QAM block for the modulation purpose. 2-PAM, 4-QAM, 16-QAM, and 64-QAM modulations are bolstered by the framework. The constellation maps for 2-PAM, 4-QAM, and 16-QAM modulation techniques are appeared in Fig. 2.13a, b, and c individually.

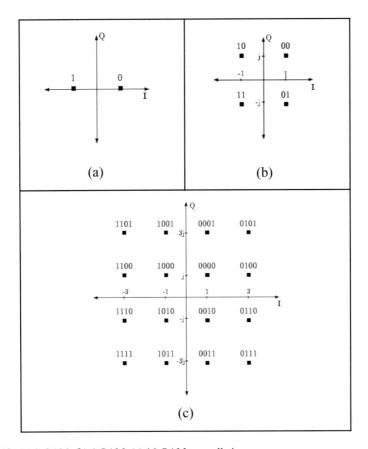

Fig. 2.13 (a) 2-QAM, (b) 4-QAM, (c) 16-QAM constellation maps

2.6.1.6   OFDM Symbols

WiMAX specifications for the 256-point FFT OFDM PHY layer characterize three kinds of subcarriers; information, pilot, and null, as appeared in Fig. 2.14. Two hundred of the complete 256 subcarriers are utilized for information and pilot sub-carriers, 8 of which are pilots forever spaced all through the OFDM range. The remaining 192 carriers take up the information subcarriers. The remainders of the potential carriers are set aside for guard bands and removal of the center frequency subcarrier. Once the information from the source is mapped into QAM symbols, the OFDM symbols must be developed. An OFDM image is made by 192 information subcarriers, 8 pilot subcarriers, 1 zero DC subcarrier, and 55 guard carriers which is equivalent to 256 that can be adequately envisioned from the inner structure of OFDM obstruct as appeared in Fig. 2.14.

It tends to be seen from the Simulink preview of OFDM internal structure which has the (192 × 1) whole numbers from the QAM have been connected to the multi-port selector block where different lines would be chosen as a type of output vectors

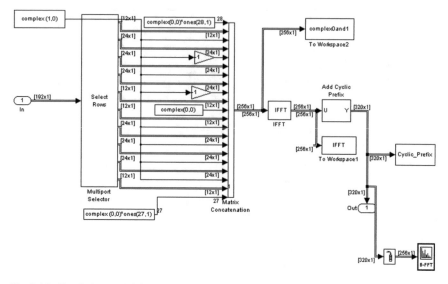

Fig. 2.14 Simulation model for generation of OFDM symbols

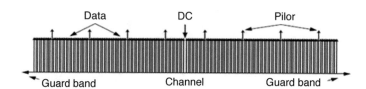

Fig. 2.15 Formation of OFDM symbols

with the measurements (24 × 1) and (12 × 1). Alongside this vector of measurement (192 × 1), some no. of guard subcarriers are embedded to the following block of network concatenator. Figure 2.15 contains the structure of OFDM symbols where (28 × 1) lower frequency guard subcarriers and (27 × 1) higher frequency subcarriers are included with information bits. Eventually the structure of OFDM symbol is the mix of information carriers, guard bands, and DC just as pilot carriers.

So as to develop an OFDM symbol, a procedure to revise these carriers is required. With this reason, the assembler block is embedded in the simulator. It completes this task in two stages by first embedding the pilot tones and the zero DC subcarrier between information with a procedure of vertical concatenation, and after that affixing the training symbols towards the start of each burst in a horizontal manner, as appeared in Fig. 2.16.

The procedure referenced as the arrangement of OFDM symbol has been executed in this given work. Here the block matrix is with 21 contributions with dimension of (256 × 1) that incorporates 192 information subcarriers, 8 pilot carrier, 1 dc subcarrier and remaining 55 are zero subcarriers annexed towards the end of the

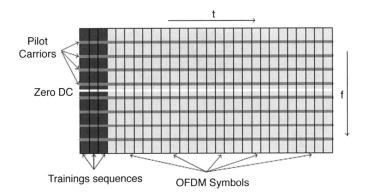

Fig. 2.16 Rearranged OFDM burst structure

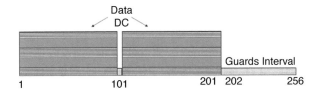

Fig. 2.17 OFDM symbol structure with shuffled guard interval

cited structure, and guard bands are also inserted to enable the natural decay of the signal. The entire structure is demonstrated in Fig. 2.17.

Here the concatenation is picked to be 1 in light of the fact that the input arrays are getting concatenated vertically. For horizontal concatenation, it needs to pick 2. The (256 × 1) subcarriers are given to the IFFT block which is utilized to deliver a time domain signal, as the symbols acquired after modulation can be viewed as the amplitudes of a specific scope of sinusoids. This implies that every one of the discrete samples before applying the IFFT algorithm relates to an individual subcarrier. Besides ensuring the orthogonality of the OFDM subcarriers, the IFFT shows to a rapid path for modulating these subcarriers in parallel, and along these lines, the utilization of multiple modulators and demodulators which spend a lot of time and resources to perform this operation is avoided. Before doing the IFFT operation in the test system, i.e., simulator, the subcarriers are rearranged. For this introduced work, the 256-point IFFT has been picked by the gotten information subcarriers. Figure 2.18 demonstrates the subcarrier structure that enters the IFFT block after performing the above-mentioned rearrangement. As found in Fig. 2.18, zero subcarriers are kept in the focal point of the structure.

The output vector of the IFFT block has been connected to the "U-Y selector" block which chooses the approaching subcarrier sequence and includes some additional stream that is known as a cyclic prefix. The vigor of any OFDM transmission

Fig. 2.18 OFDM symbol with rearranged guard band

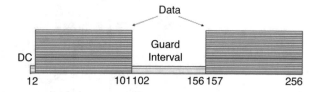

Fig. 2.19 Cyclic prefix

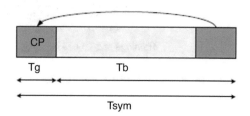

against multipath delay spread is accomplished by having a long symbol period to minimize the ISI. Figure 2.19 delineates one approach to perform the cited long symbol period, making a cyclically extended guard interval where each OFDM symbol is preceded by a periodic extension of the signal itself. This guard interval, that is really a duplicate of the last portion of the information symbol, is known as the cyclic prefix (CP).

Copying the end of a symbol and appending it to the start results in a longer symbol time. Hence, the total length of the symbol is

$$T_{\text{sym}} = T_b + T_g \tag{2.3}$$

In above equation, T_{sym} is the OFDM symbol time, where T_b is the useful symbol time and T_g shows the CP time.

The parameter G characterizes the proportion of the CP length to the helpful symbol time. When wiping out ISI, it must be considered that the CP must be longer than the dispersion of the channel. In addition, it ought to be as small as conceivable since it costs vitality to the transmitter. Hence, G is generally under 1/4 which can be justified from the index vector choice of the model parameter. As it very well may be seen from the snapshot preview of the "U-Y selector," the input data dimension is (256×1) and the output data dimension is (320×1) which shows the expansion of 64 bits as a cyclic prefix. This can be set in the "index vector" parameter of the "U-Y selector" block. Consequently, it very well may be said that with respect to an effective data stream of (256×1), the cyclic prefix amount is 64, for example, ¼ of the effective data. The OFDM modulated (320×1) subcarriers have been essentially transmitted through the communication medium signified as "WiMAX Channel" in this model, which has been examined in the following subsection.

2.6.2 *WiMAX Channel*

The most significant unit of any communication framework is the link layer that associates transmitter and receiver unit, for example, wireless communication channel. In wireless communication framework channel assumes the most basic role that is the reason why modeling of channel and decision of appropriate channel is the exceedingly attractive assignment. For the situation where transmitter and receiver both are effective however the natural condition is not ideal then the productive capacity cannot be performed by the framework. Along these lines, to build up the model for such a communication framework, the prime objective ought to be towards determination and modeling of the wireless communication channel.

Here, the fundamental modeling of the WiMAX framework has been finished by considering the ideal AWGN channel which is appeared in Fig. 2.20 by expecting that the signal to noise ratio stays constant all through the long channel. This is the genuine piece of the WiMAX modeling whereby setting the various estimations of required channel SNR, the impact of BER can be determined. The significant development of this examination work is the varieties in the determination of channel types just as modeling of multipath structure inside the channel in a manner to improve the system capacity as well as the reduction in bit error rate. Figure 2.21 demonstrates is the snapshot of the function block showing the channel parameters?

The output of the WiMAX system channel is the similar no. of subcarriers, i.e., (320 × 1), which are received by the multiple blocks of the WiMAX system receiver. The subsequent subsection details the WiMAX receiver blocks.

Fig. 2.20 Simulation model for WiMAX channel

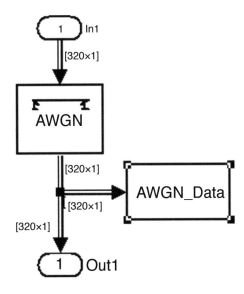

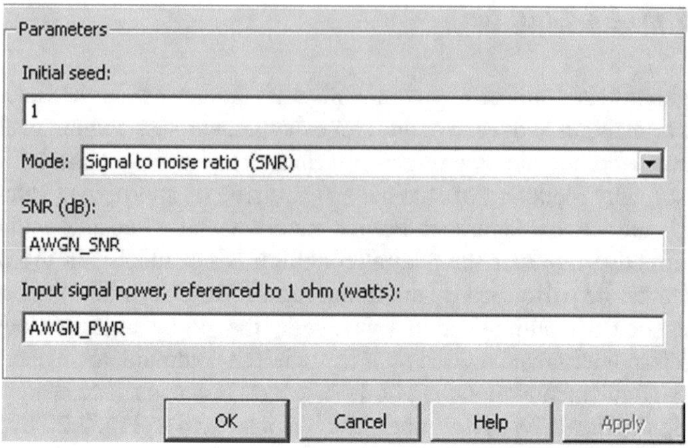

Parameters
Initial seed:
1
Mode: Signal to noise ratio (SNR)
SNR (dB):
AWGN_SNR
Input signal power, referenced to 1 ohm (watts):
AWGN_PWR
OK
Cancel
Help
Apply

Fig. 2.21 WiMAX channel parameters

De-Randomizer
RS Decoder
Viterbi Decoder
QAM Demodulator
OFDM Demodulator
Output Data
DATA RX

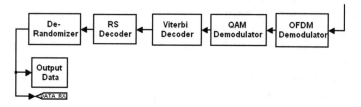

Fig. 2.22 Block diagram of WiMAX receiver

2.6.3 WiMAX Receiver

The WiMAX receiver essentially carries out the reverse operation as the transmitter as well as channel estimation necessary to reveal the unknown channel coefficients. This section usually carries out the reverse task to reconstruct the transmitted information bits. Figure 2.22 shows the snapshot of the WiMAX receiver. Further, the subsequent subsection indicates the block by block explanation.

2.6.3.1 OFDM Demodulator

The modulated and coded data got from the WiMAX channel are received through the reception antenna with or without diversity over which the OFDM demodulation would be performed. The primary block of OFDM demodulator is FFT. Figure 2.23 demonstrates the fundamental structure of the OFDM demodulator.

The precise process which is done in OFDM demodulator have been performed in this segment example

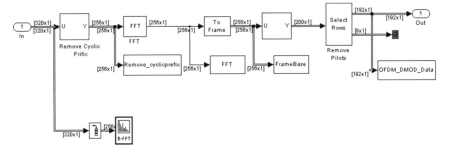

Fig. 2.23 Simulation model for OFDM demodulator

- Removal of cyclic prefix from the captured information
- Fast Fourier Transform of the time domain information
- Framing and reordering of information
- Seperation of 56 zero subcarriers from the received 256 input data that were embedded as guard bits at the transmitter side by "U-Y' selector" block
- Separation of 8 pilot carriers and 192 information subcarriers from the received 200 subcarriers

The referenced arrangement of errands has been performed by numerous sub-blocks of OFDM demodulator as showed in the snapshot. The 320 subcarriers are the generated by combining various carrier like 192 information subcarriers, 56 zero padding subcarriers, 8 pilot carriers, and 64 carriers of the cyclic prefix. The underlying "U-Y' selector" block essentially expels 64 cyclic prefix carriers and creates 256 double integers from 320 approaching carriers. These are given to FFT block basically to get the reverse fuction IFFT for example to change over the time domanin sequence to frequency domain. From this arrangement, the reframing and reordering would be performed by zero padding known as zero subcarriers for guard band reason. Out of this 200 subcarriers, the the row selector matrix would play out the exact opposite function to that of the input side for example to 192 information subcarriers and 8 pilot carriers. These would be forwarded to next process of receiver, for example, demodulation and decoding.

2.6.3.2   QAM Demodulator

The information from OFDM demodulator would go to the demodulation procedure by experiencing quadrature amplitude demodulation block where the symbol get changed over to original bits back.From the snapshot of QAM demodulator or demapper as appeared in Fig. 2.24, it tends to be seen that the absolute reverse procedure has been performed by the numerous no. of blocks like math function for example "u bar" block, QAM demapper, and integer to bit converter.

The 192 information subcarriers got from the last block of OFDM demodulator are connected to the gain bolck whose output remains as before as 192 subcarriers.

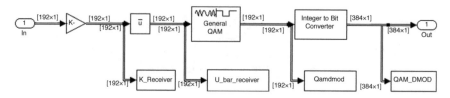

Fig. 2.24 Simulation model for QAM demodulator

As these are in the form of conjugate numbers, they ought to be connected to the math function represented as "u bar" block. From that, the 192 subcarriers are connected to the procedure of demapping where through denormalization process the 192 integer complex quantities of a structure

$$0.7071 + 0.7071i$$
$$0.7071 - 0.7071i$$
$$-0.7071 + 0.7071i$$
$$-0.7071 - 0.7071i$$

get changed into a matrix of 192 complexes number with the algebraic summation of real component and imaginary component in a form of

$$1.0000 + 1.0000i - 1.0000 + 1.0000i$$
$$1.0000 - 1.0000i - 1.0000 - 1.0000i$$

These 192 double integers as a blend of real and imaginary segments are combined as (192 × 1) whole numbers by the MATLAB directions which are finally changed over into 384 bits (for 4-QAM) by the block of integer to bit converter. These (384 × 1) bits are additionally given to the last procedure of WiMAX receiver system, for example, sequence of decoding by methods for convolution decoder and RS decoder.

2.6.3.3 Viterbi Decoder

The Viterbi decoder decodes the convolutional coded information and diminishes the computational burden by exploiting the exceptional structure of the trellis code. Another favorable position is its multifaceted nature, which is not a component of the quantity of symbols that make the codeword sequence. Figure 2.25 shows the essential design of the Viterbi decoder.

The Viterbi calculation performs approximate most extreme probability decoding. It involves calculating a measure of likeness or separation between the received signal at a time and all the trellis ways entering each state in the meantime. The algorithm works by expelling those trellis ways from the thought that could not possibly be candidates for the maximum likelihood choice. At the point when two ways

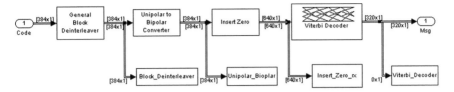

Fig. 2.25 Simulation model for Viterbi decoder

enter in same state, the one that has the best metric is picked as the "surviving" way. The determination of the distinctive "surviving" ways is performed for every one of the states. The decoder proceeds thusly to progress advance deeper into the trellis making decisions by eliminating the least likely paths. The early dismissal of unlikely path is the way that decreases the complexity. The goal of selecting the optimum path can be communicated equivalently as choosing the codeword with the maximum likelihood metric, or as choosing the codeword with the base separation metric.

Here as appeared in the depiction of Viterbi decoder, the mix of numerous blocks plays out the precise invert function to that of the convolution encoder. At the transmitter side, the "block interleaver" was the part of QAM modulator while here at the recipient side, in place of QAM demodulator, the invert of interleaving, for example, block de-interleaving has been performed by sub-block of Viterbi decoder which reorders the grouping of (384 × 1) bits. The de-interleaver revamps the bits from each burst in the right path by requesting them sequentially as before the interleaving procedure. The yield of (384 × 1) bits is given to the block of "insert zero." The block named "Insert Zeros" manages the task of reversing the procedure performed by the "Puncture" block. As recently clarified in the transmitter section, the puncturing procedure comprises of erasing bits from a stream. The receiver does not know the estimation of the erased bits but rather it can know their situation from the puncturing vectors. Along these lines, zeros are utilized to fill the relating hollows of the stream so as to get a similar code rate as before playing out the puncturing process. It again changes over approaching (384 × 1) bits into (640 × 1) bits which are connected to Viterbi decoder block for performing convolution decoding and lastly creates bit vector of size (320 × 1). Now the convolutionally decoded data is given to the external decoding by RS decoder logic and that logic has been incorporated into the accompanying subsection.

2.6.3.4 RS Decoder

The final part of the WiMAX system decoding process is the Reed–Solomon decoding shown in following Fig. 2.26. It carries out the required operations to decode the signal, and get, in the end, the original information transmitted by the transmitter source.

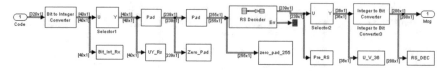

Fig. 2.26 Simulation model for RS decoder

Therefore, the RS decoder captures codewords of length n, and, after completion of decoding the signal, it proceeds messages of length k, being $n = 255$ and $k = 239$, the same as the ones mentioned in the RS encoder.

The input part of the RS decoder system of Simulink accepts vectors, with a length that are multiple of l_n. Its output is, for this situation, a vector with a length that is the same integer multiple of l_k. Subsequently, a procedure to get the right number of bytes that enter the RS decoder block, and later, revamps these bytes, should be performed initially. The structure that will enter the decoder block must be equivalent to the one that outputs the encoder block, before discarding the corresponding included bytes from the shortening and puncturing process. "pad" block deals with the procedure of rearranging the information in a matrix structure, with the predetermined size $(k + 2t) \times$ NRS, being k and t the particular parameters. The block "inserts zeros 1" includes $239 - k$ zero-bytes towards the start of the structure with the point of building the zero prefix. The $16 - 2k$ parity bytes for example (255×1) are received from second "pad" block. At long last, through "RS decoder" and "U-Y selector" block (36×1) integers are received that are changed over to (288×1) bits from integer to bit converter block which has been really produced after randomization process at the transmitter side for the transmission.

2.6.3.5 De-randomizer

This block of the receiver as shown in following Fig. 2.27 restores the original transmitted information bits back and these information bits are applied as one of the inputs of bit error rate calculator for identifing the performance of the overall WiMAX framework.

Now (288×1) bits have been passed via "U-Y selector" block to take away the additional zeros that have been padded at the WiMAX transmitter side. So the output would be (280×1) bits that have been XORed with the stream produced by PN sequence generator (same as transmitter) block to get back the original information.

2.6.3.6 Output Data

Figure 2.28 shows the basic logic of the output block. The (280×1) bits have been given to the bit to integer block to produce (35×1) integers which have been terminated right now in this traditional model of WiMAX framework. But in real-time

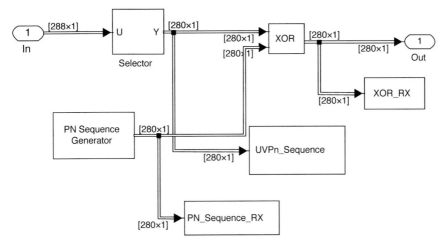

Fig. 2.27 Simulation model for de-randomizer

Fig. 2.28 Simulation
model for output data

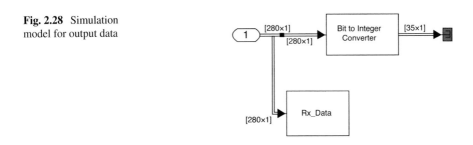

implementation of the WiMAX system, a model with real-time image transmission, output has not been terminated but has been used to retrieve the original image back.

2.6.3.7   BER Calculation

This is the block of the WiMAX model demonstrated in Fig. 2.29 whose yield determines the computations of BER values. The justification behind the realization of this block is to compare the stream of bits got from the last block of the receiver with the input bit stream and get the proportion of erroneous bits to the total no. of transmitted bits. This block takes two input information sources, one is the contribution of randomizer with (280×1) information carriers and the other is the output of de-randomizer with the equivalent no. of data, for example, (280×1). The diagrams of BER v/s referenced SNR will legitimize the framework execution regarding capacity and precision.

By considering the above examined principal parts of traditional WiMAX framework modeling, the following section of the chapter manages one of the advancements of the introduced work, for example, modeling of WiMAX transceiver

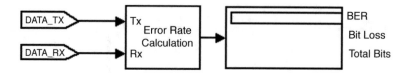

Fig. 2.29 Simulation model for BER calculator

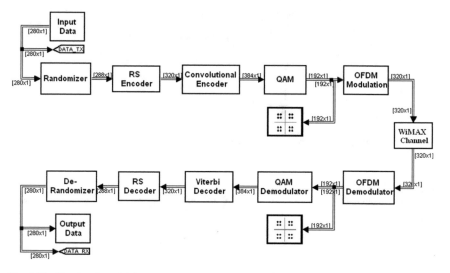

Fig. 2.30 Simulated model for traditional WiMAX system

framework utilizing real-time data in form of image transmission. Around there of wireless communication, many researchers have worked hard for the advancement and examination of the conventional model of WiMAX framework with summed up ideal channel approach. Be that as it may, this is the thing that the phase of this exhibited work gives incredible knowledge towards the real-time modeling of WiMAX framework with image transmission alongside the practical scenario of the wireless communication channel. This is the thing that the most significant advancement of this exhibited work has been expounded in the consecutive sections.

2.7 Testing of Traditional WiMAX System Model

The traditional modeling of WiMAX framework setup shown in following Fig. 2.30 includes Matlab R2009a and Communications Block set 3 running on Windows XP SP2. On the process of simulation of traditional WiMAX model by considering various system parameters as discussed in the previous sections, the following simulation outcomes have been carried out for the specific fixed values of channel signal to

Table 2.4 Parameters for testing of traditional WiMAX system model

Constant parameter	Value
RS encoder primitive polynomial	$P(x) = x^8 + x^4 + x^3 + x^2 + 1$
RS encoder generator polynomial	$g(x) = (x + \lambda^0)(x + \lambda^1)(x + \lambda^2)\ldots(x + \lambda^{2t-1})$
Convolution encoder native rate	5/6
QAM modulation order	4
N FFT	256
N-IFFT	256
Cyclic prefix	¼

noise ratio. Figure 2.30 demonstrates the snapshot of the traditional WiMAX model with random data after the MATLAB simulation. Each and every block of the whole framework has been specified with their input-output data status at individual ports.

In this case, the simulation has been performed and presented by considering the two different values of the wireless channel with SNR = 20 dB and SNR = 27 dB. Again, the rest of the other parameters have been set constant as shown in following Table 2.4.

By setting every one of these parameters, the impact of channel SNR over the framework performance is shown as the simulation results about terms of the magnitude spectrum and bit error rate with reference to total transmitted bits for the output from the transmitter. Again, by the assessment of WiMAX framework in an above referenced way, one can pass judgment on the decision of optimized block parameters for the best WiMAX framework where it is required to set the ideal estimations of square parameters for ongoing investigation by transmitting actual image rather than arbitrary information.

2.7.1   Testing of Model for Channel with SNR = 20 dB

At the fixed estimation of channel SNR = 20 dB, the straightforward examination between information and output can be assessed by methods for input and output scatter plot alongside BER calculator appeared in Figs. 2.31, 2.32, and 2.33 separately. From the scattering diagram of input information, it tends to be seen that the symbols are situated in an exceptionally precise way in each of the four quadrants as a 4-QAM technique for modulation has been picked in this model. Now as these (280 × 1) bits are transmitted through the WiMAX channel, the noise would be experienced as a real-time prototype and a portion of the bits are getting ruined by the state of the wireless channel and that can be all around seen from changes of the information bits in the diagram of output data.

Because of the wireless channel noise, justified by the significantly low value of channel SNR, the bits of all the quadrants are now no more concentrated to their original stable positions of the quadrants. They are fluctuating and getting interfered with one another that result in the bit stream errors as shown in following Fig. 2.33a, b.

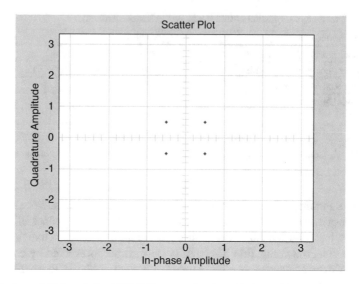

Fig. 2.31 Input QAM symbols of WiMAX system for channel with SNR = 20 dB

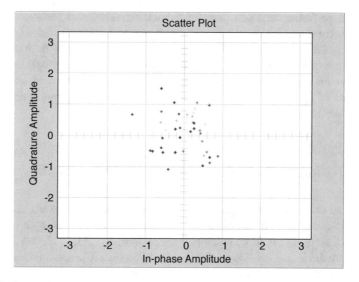

Fig. 2.32 Output QAM symbols of WiMAX system for channel with SNR = 20 dB

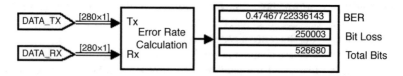

Fig. 2.33 BER calculation of WiMAX system for channel with SNR = 20 dB

The near investigation among information and out bits of the WiMAX model can be numerically demonstrated by the utility of BER calculator appeared in Fig. 2.33. As can be seen from the snapshot of the BER calculator appeared in Fig. 2.33, it very well may be seen that under 20 dB channel SNR, 250,003 bits are getting ruined while transmitting 526,680 bits. Along these lines, the BER of the WiMAX system would come around 0.47 which is nearly exceptionally huge worth and system performance can be viewed as very poor.

2.7.2   Testing of Model for Channel with SNR = 27 dB

The past subsection demonstrates that exhibition of WiMAX framework for the channel with SNR = 20 dB is exceptionally poor. Therefore, the performance of the WiMAX framework is expanded by utilizing a channel with higher channel SNR value. In this area, testing of a model for the channel with SNR = 27 dB is shown and talked about. The graphs for input information and output information appeared in Figs. 2.34, 2.35, and 2.36 have been determined by setting channel SNR = 27 dB. When contrasted with input QAM symbols, the output QAM symbols are scattered in an amount which is truly not as much as that of if there should arise an occurrence of channel SNR equivalent to 20 dB. Additionally, the estimations of BER calculator likewise legitimizes the improvement in BER as out of 526,680 bits, just 2468 bits are getting lost that outcomes in the BER of around 0.0046 as it were.

In this case, the improvement in BER is due to the fact that at very low SNR the symbols are not easy to recognize. Finally, one can conclude that the WiMAX

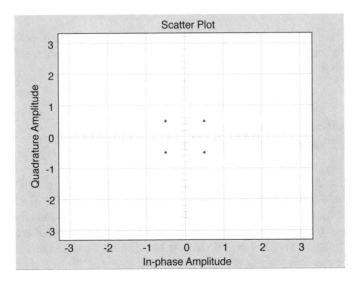

Fig. 2.34 Input QAM symbols of WiMAX system for channel with SNR = 27 dB

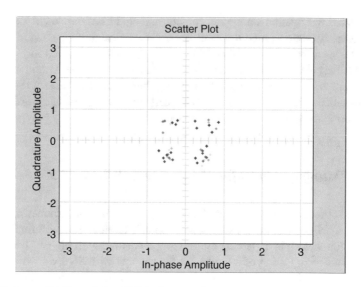

Fig. 2.35 Output QAM symbols of WiMAX system for channel with SNR = 27 dB

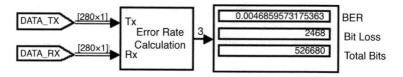

Fig. 2.36 BER calculation of WiMAX system for channel with SNR = 27 dB

system throughput is directly proportional to the SNR of the efficient wireless channel provided that the remaining parameters of the WiMAX transmitter and WiMAX receiver block must remain at the same value.

2.8 Modeling of Traditional WiMAX System for Transmission of Real-Time Image

The traditional framework for WiMAX system demonstrated in Fig. 2.37 is built on a QAM modulation technique. The modeling setup includes MATLAB R2009a and Communications Block set 3 running on Windows XP SP2. The model itself consists of three main components namely WiMAX transmitter, WiMAX receiver, and WiMAX channel. Transmitter and receiver block comprise of channel coding and modulation sub-components, whereas channel is denoted as WiMAX channel which is, for this real-time image transmission section, considered as the most generalized channel, i.e., AWGN channel.

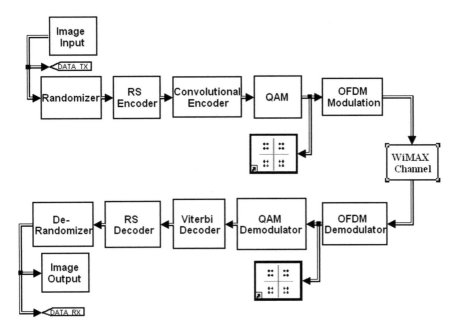

Fig. 2.37 WiMAX system model for real-time image transmission

The key difference among the traditional model of WiMAX system discussed in the earlier section and this model is the presence of two blocks, i.e., "image input" block and "image output" block, through which image information samples from MATLAB workspace are obtained at consecutive sample times and the matrix is shown in form of the image, respectively. Here as an image input 256 × 256 leaf image shown in following Fig. 2.38 has been taken whose 29 samples are successively taken at an interval of 1/29 so as to make this model compatible with the traditional model of WiMAX which passes data in matrix form only.

Figure 2.39 shows the scattered diagram of input QAM symbols that looks very decent, i.e., steady state position in their placement before transmission. After transmission through the wireless channel, these data symbols fluctuate in their position due to the channel disturbances which will be pointed by a disturbance in the output image.

As examined already in the traditional model, whatever function the random integer block was performing, here it is performed by the input information subblock of the image input block as appeared in the snapshot of Fig. 2.40. Through the MATLAB command: Image input_data = uint8(reshape(test_img, 256*256,1)) image of the measurement (256 × 256) would be fetched and sampled in terms of 29 sample consecutively which are then changed over into (232 × 1) bits by means of integer to bit converter. At that point after these (232 × 1) bits are joined with error controlling, checking and synchronization bits which will as a whole form the information of (280 × 1) bits that are additionally prepared through indistinguishable

Fig. 2.38 Input image of WiMAX system model for real-time image transmission

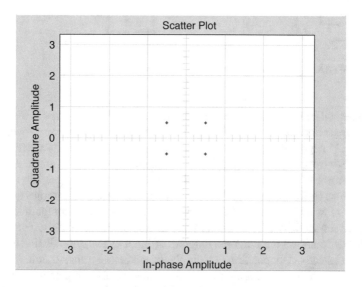

Fig. 2.39 Input QAM symbols for testing of WiMAX system

block as a traditional WiMAX model, for example, randomizer, RS encoder, convolution encoder, QAM and OFDM modulation.

Another advancement of this model is the occurrence of the block "input image viewer" for the reason of displaying the image signal. The internal structure has been shown in the following snapshot of Fig. 2.41.

In this case, the (29×1) integer samples of the image are applied to buffer of the size 256 whose output is fed to the input image viewer block where by method of two-stage matrix transposition and delay, the image of the dimension (256×256)

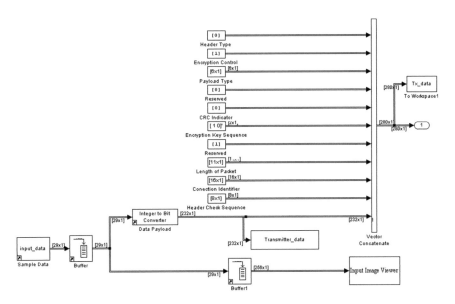

Fig. 2.40 Input image data with MAC layer standards for WiMAX system

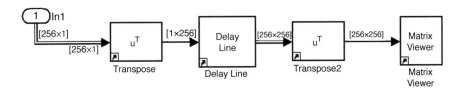

Fig. 2.41 Input image viewer

has been displayed by "matrix viewer" sub-block. To explain the MATLAB logic at the back of "image input" block, the methodology has been prepared as follows.

Program Flow
Step 1: Read the input image from a specific file.
Step 2: Resizing the image in 256×256 size matrix and send to the image in the viewer.
Step 3: Rearrange the 256×256 size matrix into $65,536 \times 1$ single array matrix for serial pixel transmission.
Step 4: At receiver, convert $65,536 \times 1$ matrix into 256×256 size matrix and send to image out the viewer.

Presently once the underlying period of image acquisition and processing on it to change over it into the form of bit is over, the data having (280×1) bits are handled by the blocks of randomizer, RS encoder, convolution encoder, QAM and OFDM modulation which is actually the replica of the procedures formed into the traditional WiMAX framework with random information instead of image signal transmission.

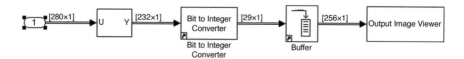

Fig. 2.42 Output data for WiMAX system for real-time image transmission

The randomizer joins the approaching (280 × 1) bits with the arbitrary PN sequence bits by methods for XORing and produces output of (288 × 1) matrix that has been gone through the cascading of RS encoder and convolution encoder as examined as of now in traditional WiMAX model whose last output is bit sequence of (384 × 1) bits. Further, the encoded information have been modulated by QAM block and connected to the procedure of OFDM modulation that at long last give output in terms of (320 × 1) integer subcarriers which are the complex no. formed by real and imaginary components that have experienced through the way towards interleaving, complex conjugation, the inclusion of a cyclic prefix, and so forth. The information of (320 × 1) would be gone through the WiMAX channel. Channel reaction on the data results in degradation of information by means errors relying upon SNR of the channel.

Usually reverse operation of transmitter is done by the receiver and signal is reconstructed again. This model is differing mainly by one block at output side as compares to traditional model, i.e., "image output" block. The rest of the blocks, i.e., OFDM demodulation, QAM demodulation, convolution decoding, RS decoding, and de-randomization, have processed over the data in the same fashion to that discussed in the traditional WiMAX model. After the above all processes, the "output image" block, the last block of the WiMAX receiver acquires (280 × 1) bits which have been further processed through the multiple blocks as explained in the snapshot Fig. 2.42.

The received incoming (280 × 1) bits are further under medication and process through "U-Y selector" block which extract the extra 48 bits that have been added at the transmitter side for security, synchronization, and error checking purpose. The out (232 × 1) bits are converted into (29 × 1) integer stream which were not terminated here as with the case of traditional WiMAX modeling with random input-output data but they have been given to buffer which gives output a matrix of (256 × 1) that has been further processed through the block of "output image viewer" to generate the original image back for the comparative analysis purpose. The internal logic of "output image viewer" block has been explained in Fig. 2.43.

Here through two-stage matrix transposition process and delay output image with the dimension of (256 × 256) has been constructed. This has been explained in terms of program flow as shown below.

Program Flow
Step 1: At receiver, convert 65,536 × 1 matrix into 256 × 256 size matrix.
Step 2: Resizing the image in 256 × 256 size matrix and send to image out the viewer.

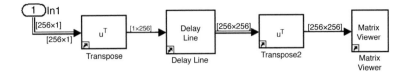

Fig. 2.43 Output image viewer

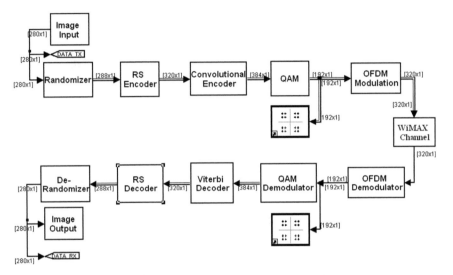

Fig. 2.44 Simulation model for traditional WiMAX system for real-time image transmission

According to the logic developed by program flow WiMAX model with the transmission of the real-time image signal has been simulated. The output image can be obtained which will be compared with the original input image signal to reduce the performance of the WiMAX system.

One of the main uniqueness of this presented work is the real-time transmission of the image rather than random data through the traditional model of WiMAX. Obtained result for the real-time data is satisfactory. Figure 2.44 shows the snapshot of the updated WiMAX model with real-time image transmission after the MATLAB simulation. Each and every block of the whole system has been specified with their input-output data status at individual ports. To evaluate the system performance in terms of BER, all the block parameters have been set at the optimum values as experimented and analyzed in the previous section except channel SNR. Here the simulation of traditional WiMAX system for real time image transmission has been performed by taking channel with SNR = 28 dB in model.

The snapshots of Figs. 2.45, 2.46, and 2.47 indicate scattering diagram of output OFDM symbols, output image, and BER calculator, respectively, of the WiMAX model for real-time image transmission. Here the 256 × 256 input image gets

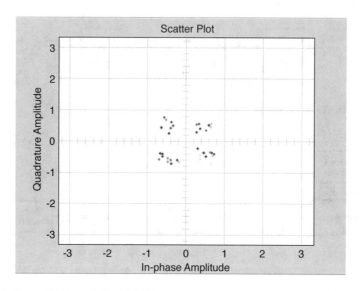

Fig. 2.45 Output QAM symbols of WiMAX system model for real-time image transmission

Fig. 2.46 Output image of WiMAX system model for real-time image transmission

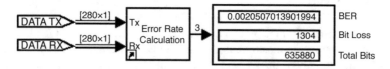

Fig. 2.47 BER calculation of WiMAX system model for real-time image transmission

converted into a data matrix of (280 × 1) and then after encountering various processes of modulation and encoding when transmitted through the WiMAX channel having SNR equal to 25 dB, the output image gets disturbances in terms of noise at the receiver side. The above phenomenon can be very well observed by comparing input and output images as well as their constellation diagrams. The fact is also supported by a reading of BER calculator where by transmitting 635,880 bits, 1304 bits are lost during transmission and because of that BER is 0.002.

It can be clearly observed by comparing the output image with the original image that by increasing channel SNR, thereby the reading of BER calculator has been improved. The total bit loss is of just 1304 bits with respect to same no. of 635,880 transmitted bits as compared to case 1.

Here again, the similar conclusion can be derived that wireless system performance is highly SNR dependent of wireless channel, i.e., what amount of disturbance applied by the channel on data which is flowing through the channel. To improve the performance of the system, higher value of SNR should be achieved by its not feasible solution every time because it can be achieved by increasing signal strength of transmitted signal. To overcome this limitation, various antenna diversity algorithms should be used in the WiMAX system with data and image at the lower value of channel SNR which has been analyzed, simulated, and proved in the next chapters.

References

1. Yarali, A., & Rahman, S. (2008, May). WiMAX Broadband Wireless Access Technology: Services, architecture and deployment models. In *2008 Canadian Conference on Electrical and Computer Engineering* (pp. 000077–000082). IEEE.
2. Saeed, R. A., Mabrouk, A. A., Mukherjee, A., Falcone, F., & Wong, K. D. (2010). WiMAX, LTE, and WiFi interworking. *Journal of Computer Networks and Communications, 2010*, 1–2.
3. Liangshan, M., & Dongyan, J. (2005). The competition and cooperation of WiMAX, WLAN and 3G. Mobile Technology, Applications & Systems. In *The 2nd International Conference*.
4. Kumar, A. (2014). *Mobile broadcasting with WiMAX: Principles, technology, and applications*. London: CRC Press.

Chapter 3
Various Techniques for WiMAX System Modeling

3.1 Introduction

The main purpose of this book is to produce the WiMAX system along with the implementation of various antenna diversity techniques coupled with Alamouti scheme in it to improve the capacity of the system without changing the bandwidth requirement of system. Nowadays no hardware is produced without positive concern of strong simulation tool on the high-speed computer.

The purpose of this chapter is to explain some steps in bridging the gaps between system and hardware level simulation supported MATLAB. The goal is to be able to directly see the impact of component utility and design along with its internal structure and its effect on system level performance measures. The motivation to describe this chapter is to design wireless communication system with the various implementation of antenna diversity systems, and also the performance of that on the given system in terms of the bit error rate (BER) or another Quality-of-Service (QoS) measure.

This chapter is principally divided into three major parts. In the initial part introduction of various diversity techniques. The second half is restricted to one of the foremost promising and widely used space diversity systems with its advantages and design to existing communication system. Last phase describes various coding and flow of implementation of various diversity techniques, i.e., single input multiple output (SIMO), multiple input single output (MISO), and multiple input multiple output (MIMO) along with its critical comparative analysis by taking many parameters under consideration has been administered so final outcome in the form of BER and system capability.

© Springer Nature Switzerland AG 2020
B. S. Sedani et al., *WiMAX Modeling: Techniques and Applications*,
https://doi.org/10.1007/978-3-030-22460-8_3

3.2   Diversity Techniques

In wireless communication, radio waves traveling along different paths arrive at the receiver at different times with random phases and combine constructively or destructively as shown in Fig. 3.1.

In the wireless link, different signal will travel with different type of channels. Some channel forms direct line of sight path, and other path components are due to various obstacles in the channel. Because of that obstacles, signal needs to travel through the various paths towards the receiver. Because of various paths in the wireless channel, delay is different. Different delay in the component generates phase delay among them. Vector summation would be carried out to find out resultant signal at the receiver. Figure 3.2 shows vector combination of that.

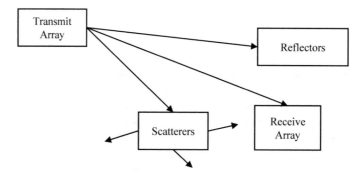

Fig. 3.1 Mechanism for wave propagation

Fig. 3.2 Multipath components of wave propagation

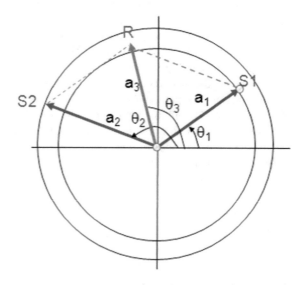

When two or more multipath components are arrived at the receiver with the same access delay at the same time, the received signal is the vectorial addition of two multipath signals. Let's assume that two signals S_1 and S_2 arrive at the same time at the receiver and R is the combined signal at the receiver.

$$S_1 = a_1 e^{j\theta_1}, \quad S_2 = a_2 e^{j\theta_2}$$
$$R = S_1 + S_2 = a_1 e^{j\theta_1} + a_2 e^{j\theta_2}$$

(3.1)

The actual outcome is a frequent variation in the amplitude of the received signal in a very small period of time or distance traveled known as fading. At the same time, the large-scale average path loss leftovers constant. Diversity takes advantage of multipath propagation to strengthen the signal. Conventionally, the design of wireless systems has been focused on increasing the reliability of the air interface; in this context, fading and interference are viewed as nuisances that are to be countered. Recent focus has shifted more towards increasing the spectral efficiency; associated with this shift is a new point of view that fading can be viewed as an opportunity to be exploited.

While dealing with the multipath environment, the individual signal path arriving at the receiver faces independent or highly uncorrelated fading. This means that when a particular signal path is in a fade there may be another signal path not in any fade. This phenomenon of independent fading in various paths can be exploited as an advantage to achieve improved performance in wireless communication provided that out of multiple paths, at least one path can be obtained with minimum distortion and maximum signal strength. This phenomenon leads towards the concept of diversity which can dramatically improve the performance over fading channels.

In practice, diversity systems can be applied in space, frequency, or time domains. Diversity over time can be obtained via coding and interleaving: information is coded and the coded symbols are dispersed over time in different coherence periods, so that different parts of the codewords experience independent fades. Analogously, one can also exploit diversity over frequency if the channel is frequency selective. In a channel with multiple transmit or receive antennas spaced sufficiently far enough, diversity can be obtained over space as well. In a cellular network, macro-diversity can be exploited by the fact that the signal from a mobile can be received at two base stations. Since diversity is such an important resource, a wireless system typically uses several types of diversity. The following subsections illustrate the various types of diversity systems.

3.2.1 Time Diversity

Time diversity means transmitting identical messages in different time slots as shown in Fig. 3.3. This yields two unrelated signals at the receiving finish. A similar information is repeatedly transmitted at totally different time slots with the hope that they are going to suffer fading effect in individual channel at different level and therefore the receiver will club them properly to generate the strengthened signal at last.

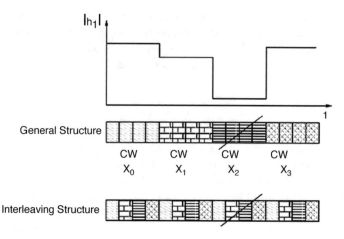

Fig. 3.3 An example of time diversity

Time division multiple access schemes are relied on the principal of time diversity which is widely used in GSM structure. Frequency division duplex system is used in GSM and uses two 25 MHz bands. Forward as well as reversed channel uses 25 MHz for assigning channels. The GSM bands are at 890–915 MHz (uplink) and at 935–960 MHz (downlink). Each traffic channel uses 200 KHz for data transmission and each channel is divided into 8 time slots in time division fashion. The data of each user are sent over time slots of length 577 µs and the time slots of the 8 users together form a frame of length 4.615 ms [1]. GSM generates 20 ms frame of encoded data. Other data is processed by convolution encoder with ½ rate and finally 456 bits are prepared at different level of encoding. To get benefit of time diversity, coded bits are rearranged and interleaved across eight time slots given to that user in specified manner. Graphically process is explained in Fig. 3.3.

For simplicity, let us consider a flat fading channel has been taken. Transmitted data is codeword $x = [x_1 \ldots x_L]^t$ of length L symbols and the received signal is given by:

$$y_l = h_l \cdot x_l + w_l \tag{3.2}$$

where $l = 1, 2, \ldots, L$.

Ideal interleaving is considered so that consecutive symbols x' are transmitted sufficiently far apart in time, and it can be assumed that the h_l's are not dependent. L parameter can be identified as numbers of diversity branches. $w_0, \ldots, w_l$ are random variables and are taken as the additive noises.

Main motive of block interleaving is to reduce burst errors occurred while transmission of signal in noisy channel. In the WiMAX system also, this type of block interleaving scheme is used to improve signal quality after reception. In the previous chapter, it has been demonstrated that physical layer of WiMAX uses this kind of interleaving and time diversity advantages can be taken perfectly.

3.2.2 Frequency Diversity

To analyze the concept of frequency diversity, consider first the one-shot communication situation when one symbol $x[0]$ is sent at time $= 0$, and no symbols are transmitted after that. The receiver observes

$$y[l] = h_l[l] \cdot x[0] + w[l] \tag{3.3}$$

where $l = 0, 1, 3, \ldots, L$.

Channel has been assumed with response of finite number of taps L, then the delayed replicas of the signal are providing L branches of diversity in detecting $x[0]$ since the tap gains $h_l[l]$ are assumed to be independent. This diversity is known as frequency diversity as this diversity is achieved by the ability of resolving the multipaths at the receiver due to the wideband nature of the channel. A straightforward communication scheme can be established on the above concept by sending an information symbol every L symbol time. The highest diversity gain of L can be obtained at a cost of transmitting one symbol at every delay spread but this scheme results in wasteful degree of freedom. This scheme is analogous to the repetition of codes used for both time and spatial diversity purpose in which one information symbol is repeated L times. In this arrangement if symbols are transmitted more frequently then intersymbol interference occurs. Now the issue is how to deal with intersymbol interchange while at the same time exploiting the inherent frequency diversity in the channel. Mainly three common approaches to deal with mentioned problem are per as below:

- **Single Carrier Systems with Equalization**: ISI can be controlled at some level by using linear and nonlinear processing at the receiver. Optimal ML detection of the transmitted symbols can be implemented using the Viterbi algorithm. Viterbi algorithm only can be used when a smaller number of taps are there in the system else it increases exponentially. Alternatively, linear equalizers attempt to detect the current symbol while linearly suppressing the interference from the other symbols.
- **Direct Sequence Spread Spectrum**: This method represents that information symbols are modulated by a pseudo-noise sequence and transmitted over a bandwidth W much larger than the data rate. As the symbol rate is very low, ISI can be reduced and receiver structure can be simpler. Although this leads to an inefficient utilization of the total degrees of freedom in the system from the perspective of one user, this scheme allows multiple users to share the total degrees of freedom, with users appearing as pseudo noise to each other.
- **Multi-carrier Systems**: In this method, transmit pre-coding is prepared to convert the ISI channel into a set of non-interfering, orthogonal subcarriers, each experiencing narrowband flat fading. Diversity can be obtained by coding across the symbols in different sub-carriers. This method is also called Discrete Multi-Tone (DMT) or Orthogonal Frequency Division Multiplexing (OFDM).

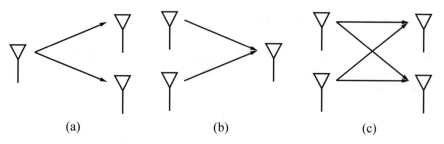

Fig. 3.4 Antenna diversity systems (**a**) SIMO, (**b**) MISO, (**c**) MIMO

3.2.3   Antenna Diversity

To implement time diversity, interleaving and coding of data set over several coherence time periods are necessary. When there is a specified delay constraint and/or the coherence time is large, this may not be possible. In this case, other forms of diversity, i.e., antenna diversity or space diversity, have to be obtained. Figure 3.4a–c show various types of antenna diversity systems. Two kinds of space diversity help to improve system and its capacity named T_x-Diversity and R_x-Diversity. T_x-Diversity uses multiple transmit antennas which are implemented for the signal transmission which in turn results in Multiple Input Single Output (MISO) diversity ($n \times 1$ system) while R_x-Diversity uses multiple receive antennas which are implemented for the signal reception which in turn results in Single Input Multiple Output (SIMO) ($1 \times n$ system). Channels with multiple transmit and multiple receive antennas so-called Multiple Input Multiple Output (MIMO) ($n \times n$) channels provide even more potential.

Antenna diversity, or spatial diversity, can be obtained by placing multiple antennas at the transmitter and/or the receiver. The antenna separation is decided by local scattering environment along with carrier frequency values. In most of the cases, mobiles are nearer to ground level with many scatters around, so that the channel de-correlates over shorter spatial distances, and typical antenna separation of the half to one carrier wavelength is proper. Larger antenna separation of several to 10's of wavelengths is required for the base stations on high tower. Following section describes various features and parameters of antenna diversity system with critical performance analysis point of view in context with real environment.

3.3   Various Antenna Diversity Systems

The idea behind antenna diversity is that if the antennas are spaced sufficiently way apart, they fade severally. By invariably choosing the antenna with the most effective channel, or (better) combining the two with applicable weights, the chance of poor reception (signal outage) is dramatically reduced. Diversity will increase the average amplitude, which successively improves capability. Though the capability increase is considerably less with diversity than if spatial multiplexing was used, it is, in general, additional strong and may be used at lower signal to noise ratios. A very efficient approach for achieving space diversity at the transmitter, while not knowing response of the channel at the transmitter, is termed space-time coding, i.e., Alamouti coding. Before approaching the various antenna diversity systems, it is quite necessary to grasp the normal antenna systems together with its limitations.

3.3.1   Single Input Single Output (SISO) System

Figure 3.5 shows the conventional antenna system with single transmitter and single receiver antenna known as SISO system. The advanced antenna arrangement is based on the principle of antenna diversity. In the starting stages, the different modulation schemes like coherent BPS, coherent QPSK, coherent 4-PAM, and coherent 16 QAM were there in which error probability reduces very slowly and is proportional to 1/SNR. All mentioned techniques were not used the principle of antenna diversity. They all were single antenna system at both end of communication system. However, diversity can be taken in such situation by implementing the OFDM technique in the form of frequency diversity while transmitting symbols. In these techniques, poor performance may be there as strength of the signal is decided by one path only.

Basically, it is experimented that bit error rate (BER) performance of such a system is decided only on channel SNRs. In this situation, there is a significant probability that taken path may be affected with highest amount of fading and interference. Such systems can be designed better by introducing a greater number of antennas in the transmitting side and/or receiver side. As Shannon theorem says, in any given channel corrupted by an additive white Gaussian noise at a level of SNR the capacity can be calculated as follows.

$$C = B \cdot \log_2 \left[1 + \text{SNR} \right] \left(\text{bps} / \text{Hz} \right) \tag{3.4}$$

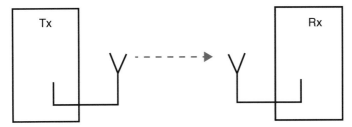

Fig. 3.5 Single Input Single Output (SISO) antenna system

where C is the Shannon limits on channel capacity, SNR is signal to noise ratio, and B is the bandwidth of the channel.

It can be concluded that theoretically capacity increases as the bandwidth or SNR is increased which is not a feasible case beyond certain limits. Therefore, to increase the capacity, a diversity mechanism is implemented on the transmitter and receiver side. The following sections illustrate the same mechanism [1].

3.3.2 Single Input Multiple Output (SIMO) System

The basic step towards diversity is to implement single input multiple output (SIMO) system configuration, which implies one transmit and two receive antennas. This is shown in Fig. 3.6. For example, a base station with one transmit and two receive antennas would be a SIMO system ($1 \times N$ system).

In a flat fading channel with 1 transmit antenna and 2 (N) receive antennas, the channel model is as follows:

$$y_l\left[m\right] = h_l\left[m\right] \cdot x\left[m\right] + w_l\left[m\right] \tag{3.5}$$

where $l = 1, 2, ..., N$, and the noise $w_l[m]$ is independent across the antennas.

System would like to detect $x[1]$ based on $y_l[1]$, ... $y_N[1]$. If the antennas are spaced sufficiently far apart, then it can be assumed that the gains $h_l[1]$ are independent Rayleigh, and we get a diversity gain of N. For SIMO system, with $N = 2$ receiving antennas for our case, the channel capacity can be given by:

$$C_{SIMO} = B \cdot \log_2\left[1 + N \cdot SNR\right]\left(bps / Hz\right) \tag{3.6}$$

where the value of SNR will be increased by factor L, i.e., $SNR = L \times SNR$.

Main advantage is that capacity has been improved and it is not only decided by SNR but also by the number of antennas. By this way advantage of diversity can be taken. SIMO performance is better than SISO.

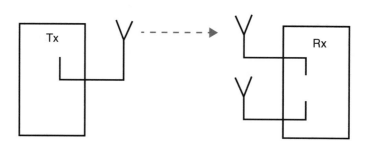

Fig. 3.6 Single Input Multiple Output (SIMO) antenna system

3.3.3   *Multiple Input Single Output (MISO) System*

One more step towards diversity principle is to implement two transmit antennas and one receive antenna. This configuration is called multiple inputs single output (MISO) system which is shown in Fig. 3.7. For example, a base station with two transmit and one receive antennas would be MISO ($M \times 1$ system).

M transmit antennas and one receive antenna is common to all downlink of a cellular system since it is often cheaper to have multiple antennas at the base station than to having multiple antennas at every handset. It is easy to get a diversity gain of M: simply transmit the same symbol over the M different antennas during M symbol times. The channel capacity of the MISO system is given by:

$$C_{\mathrm{MISO}} = B \cdot \log_2 \left[1 + M \cdot \mathrm{SNR} \right] \left(\mathrm{bps} / \mathrm{Hz} \right) \qquad (3.7)$$

where again SNR increases by the factor M which is now number of transmitting antennas of the system. Compared to the SISO system, the capacity of SIMO and MISO system shows improvement. The increase in capacity is due to the spatial diversity which reduces fading and SNR improvement. However, the SNR improvement is limited, since the SNR is increasing inside the log function.

To increase the capacity further by increasing the number of antennas within the same physical dimension, the transmit diversity scheme can be imposed for channel coding. Space-time code is the most suitable solution proposed by Alamouti, which was known as the Alamouti scheme [2]. This is the transmit diversity scheme proposed in several third-generation (3G) cellular standards. Alamouti scheme is designed for 2 transmit antennas. Here the detail of the Alamouti coding scheme with MISO antenna diversity has been discussed.

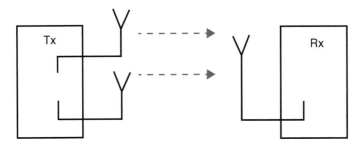

Fig. 3.7 Multiple Input Single Output (MISO) antenna system

3.3.3.1   Mathematical Modeling of MISO System Using Alamouti Coding Scheme

Most efficient approach to exploit diversity is space-time coding (STC) given by the Multiple Input Single Output and Multiple Input Multiple Output. It is used to obtain gains due to spatial diversity via multiple transmit and receive antennas. One popular representation of these codes is the Alamouti scheme for two transmit antennas. STC coding is used to improve the performance of the MISO system. Main focus is the utilization of multipath effects in order to achieve very high spectral efficiencies. With this motive, the principal aim of the space-time coding lies in the design of 2D signal matrices which is prepared for the transmission during a specified time period on a number of antennas. By this way, redundancy in space can be increased through the addition of multiple antennas, enabling us to exploit diversity in the spatial dimension, as well as obtaining a coding gain. Therefore, the transmit diversity plays an integral role in the STC design. Alamouti introduced a very simple scheme of space-time block coding (STBC) allowing transmissions from two antennas with the same data rate as on a single antenna [3].

The Alamouti algorithm uses the space and the time domain to encode data as shown in Fig. 3.8, the Alamouti algorithm uses the space and the time domain to encode data, increasing the performance of the system by coding the signals over the different transmitter branches. Thus, the Alamouti code achieves high diversity gain with full data rate as it transmits two symbols in two-time intervals.

In the first time slot, transmit antennas T_{x1} and T_{x2} are sending symbols s_0 and s_1, respectively. In the next time slot, symbols $-s_1^*$ and s_0^* are transmitted, where $(\cdot)^*$ denotes complex conjugation. Each symbol is multiplied by a factor of a squared root of 2 in order to achieve a transmitted average power of 1 in each time step. Furthermore, it is supposed that the channel, which has transmission coefficients h_1 and h_2, remains constant and frequency flat over the two consecutive time steps. The received vector, r, is formed by stacking two consecutive received data samples in time, resulting in

$$r = \frac{1}{\sqrt{2}} Sh + n \qquad (3.8)$$

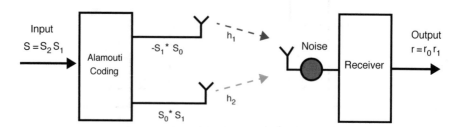

Fig. 3.8 Mathematical model of MISO System

where $r = [r_0, r_1]^T$ represents the received vector, $h = [h_0, h_1]^T$ is the complex channel vector, $n = [n_0, n_1]^T$ is the noise at the receiver, and S defines the STC:

$$S = \begin{pmatrix} S_0 & S_1 \\ S_1^* & -S_0^* \end{pmatrix} \tag{3.9}$$

The vector equation in Eq. (3.8) can be explicit as

$$r_0 = \frac{1}{\sqrt{2}} S_0 \cdot h_1 + \frac{1}{\sqrt{2}} S_1 \cdot h_2 + n_0 \tag{3.10}$$

$$r_1 = \frac{-1}{\sqrt{2}} S_1^* \cdot h_1 + \frac{1}{\sqrt{2}} S_0^* \cdot h_2 + n_1 \tag{3.11}$$

At the receiver, the vector y of the received signal is formed according to $y = [r_0, r_1]^T$, which is equivalent to

$$r_0 = \frac{1}{\sqrt{2}} S_0 \cdot h_1 + \frac{1}{\sqrt{2}} S_1 \cdot h_2 + n_0 \tag{3.12}$$

$$r_2^* = \frac{1}{\sqrt{2}} S_0 \cdot h_2^* - \frac{1}{\sqrt{2}} S_1 \cdot h_1^* + n_1^* \tag{3.13}$$

These both Eqs. (3.11) and (3.13) can be rewritten in a matrix system as specified in the below equation.

$$\begin{pmatrix} r_1 \\ r_2^* \end{pmatrix} = \frac{1}{\sqrt{2}} \begin{pmatrix} h_1 & h_2 \\ h_2^* & -h_1^* \end{pmatrix} \begin{pmatrix} S_0 \\ S_1 \end{pmatrix} + \begin{pmatrix} n_0 \\ n_1^* \end{pmatrix} \tag{3.14}$$

The Hermitian of the virtual channel matrix is

$$H_v^H = \begin{pmatrix} h_1^* & h_2 \\ h_2^* & -h_1 \end{pmatrix} \tag{3.15}$$

Finally, the estimated transmitted signal is given by $\hat{S} = H_v^H \cdot y$ and therefore,

$$\begin{pmatrix} \hat{S}_0 \\ S_1 \end{pmatrix} = H_v^H \begin{pmatrix} r_0 \\ r_1^* \end{pmatrix} \tag{3.16}$$

$$\begin{pmatrix} \hat{S}_0 \\ S_1 \end{pmatrix} = H_v^H \cdot H_v \begin{pmatrix} S_0 \\ S_1 \end{pmatrix} + H_v \begin{pmatrix} n_0 \\ n_1^* \end{pmatrix}$$

$$\begin{pmatrix} \hat{S}_0 \\ S_1 \end{pmatrix} = \frac{1}{\sqrt{2}} \begin{pmatrix} h_1^* & h_2 \\ h_2^* & -h_1 \end{pmatrix} \cdot \begin{pmatrix} h_1 & h_2 \\ h_2^* & -h_1^* \end{pmatrix} \cdot \begin{pmatrix} S_0 \\ S_1 \end{pmatrix} + \begin{pmatrix} h_1^* & h_2 \\ h_2^* & -h_1 \end{pmatrix} \cdot \begin{pmatrix} n_0 \\ n_1^* \end{pmatrix}$$

$$\begin{pmatrix} \hat{S}_0 \\ S_1 \end{pmatrix} = \frac{1}{\sqrt{2}} \left(|h_1|^2 + |h_2|^2 \right) \begin{pmatrix} 1 & 0 \\ 0 & 1 \end{pmatrix} \begin{pmatrix} S_0 \\ S_1 \end{pmatrix} \begin{pmatrix} h_1^* n_0 + h_2 n_1^* \\ h_2^* n_0 - h_1 n_1^* \end{pmatrix}$$

$$\begin{pmatrix} \hat{S}_0 \\ S_1 \end{pmatrix} = \frac{1}{\sqrt{2}} h^2 I_2 S + \tilde{n}$$

(3.17)

Once the corresponding operations for estimating the transmitted signal have been performed, the result is represented in Eq. (3.17), where

- $h^2 = |h_1|^2 + |h_2|^2$ is the power gain of the channel.
- I_2 is the 2×2 identity matrix.
- $S = [S_0, S_1]^T$ represents the transmitted symbols.
- $\tilde{n} = \begin{pmatrix} h_1^* n_0 + h_2 n_1^* \\ h_2^* n_0 - h_1 n_1^* \end{pmatrix}$ is some modified noise.

3.3.4 Multiple Input Multiple Output (MIMO) System

MIMO introduces the transmission of two streams using two or more than two spatially separated antennas. The streams are received at the receiver by using spatially separated antennas. The streams are then separated by using space-time processing, which forms the core of the MIMO technology. A base station using two transmit antennas and two receive antennas is referred to as MIMO ($n \times n$). Figure 3.9 shows the schematic of the MIMO system.

The main attraction of MIMO channels over SISO channels is the various gains like array, diversity, and the multiplexing. Array gain and diversity gain are also associated with SIMO and MISO channels [3]. Multiplexing gain, however, is a

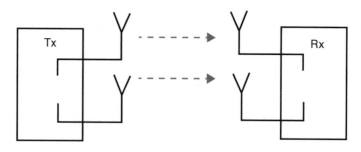

Fig. 3.9 Multiple Input Multiple Output (MIMO) antenna system

unique characteristic of MIMO channels. Array gain is the improvement in SNR obtained by coherently combining the signals on multiple transmit or multiple receive dimensions and is easily characterized as a shift of the BER curve due to the gain in SNR. Diversity gain is the improvement in link reliability obtained by receiving replicas of the information signal through independently fading links, branches, or dimensions. It is characterized by a steeper slope of the BER curve in the low BER region. As discussed in MISO, the improvement of BER can be achieved using 2×1 Alamouti coding scheme, multiple input multiple output system also exploits the averaging at receiver (Like 1×2 SIMO) and hence the dramatical improvement of BER performance can be achieved. Basically, two kinds of MIMO diversity schemes have been identified that are discussed below.

3.3.4.1 Matrix *A* MIMO System

One technique to use 2×2 MIMO is to send identical data streams on both the transmit antennas and use space time coding techniques (STC) to take advantage of the space and time diversity achieved. The effective SNR seen by the receiver can be improved by using STC with 2×2 and thus permits the use of the highest modulation coding with relatively low FEC. This effectively increases the data transmission rate. This mode of operation using space-time coding is called MIMO Matrix *A* which has been represented by Fig. 3.10.

3.3.4.2 Matrix *B* MIMO System

In an elevated SINR environment, the two transmit antennas can carry independent data streams by using a technique called spatial multiplexing (SM). Thus, each of the two streams and the peak data rate handled over the physical layer can go up to nearly double of a single stream in ideal transmission conditions. The transmission rate is very high nearer to 50% improvement than a single transmitting antenna even in characteristic field conditions. This technique of using MIMO (i.e., by using spatial multiplexing) is called MIMO Matrix *B* that can be viewed from Fig. 3.11.

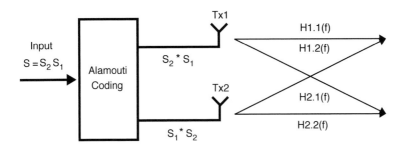

Fig. 3.10 Matrix *A* MIMO system

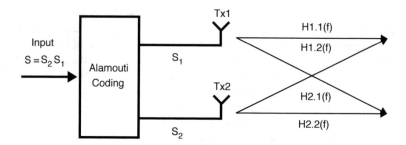

Fig. 3.11 Matrix *B* MIMO System

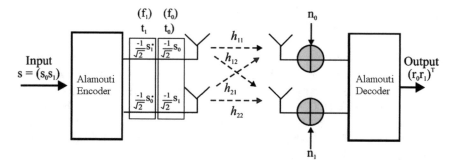

Fig. 3.12 Mathematical model for MIMO system

In this book, Matrix *A* MIMO system model is being implemented for the critical performance demonstration of transmitter and receiver diversity system and for the implementation of the antenna diversity system in the WiMAX system.

3.3.4.3 Mathematical Model for MIMO System Using Alamouti Coding Scheme

As per the theoretical aspect of MIMO given in above sections, one mathematical model of the MIMO system has been developed and can be visualized from the inspection of Fig. 3.12.

The received signal from a 2 × 2 Alamouti scheme as depicted from Fig. 3.12 is

$$
y = \begin{pmatrix} r_0(1) \\ r_0(2) \\ r_1^*(1) \\ r_1^*(2) \end{pmatrix} = \frac{1}{\sqrt{2}} \begin{pmatrix} h_{11} & h_{21} \\ h_{12} & h_{22} \\ h_{21}^* & -h_{11}^* \\ h_{22}^* & -h_{12}^* \end{pmatrix} \begin{pmatrix} S_0 \\ S_1 \end{pmatrix} + \begin{pmatrix} n_0(1) \\ n_0(2) \\ n_1^*(1) \\ n_1^*(1) \end{pmatrix} \tag{3.18}
$$

The estimated transmitted signal can be calculated from $\tilde{S} = H_Y^H \cdot y$, where $y = \left[r_0(1) \, r_0(2) \, r_1^*(1) \, r_1^*(2) \right]^T$. The virtual channel matrix H_v is expressed as,

$$H_v = \begin{pmatrix} h_{11} & h_{21} \\ h_{12} & h_{22} \\ h_{21}^* & -h_{11}^* \\ h_{22}^* & -h_{12}^* \end{pmatrix} \tag{3.19}$$

Therefore, the Hermitian of the virtual channel matrix is

$$H_Y^H = \begin{pmatrix} h_{11}^* & h_{12}^* & h_{21} & h_{22} \\ h_{21}^* & h_{22}^* & -h_{11} & -h_{12} \end{pmatrix} \tag{3.20}$$

The estimation of the transmitted symbols is performed as follows:

$$\begin{pmatrix} \hat{S}_0 \\ \hat{S}_1 \end{pmatrix} = \frac{1}{\sqrt{2}} h^2 I_2 S + \tilde{n} \tag{3.21}$$

Equation (3.21) expresses the obtained result for the process of estimating the transmitted symbols.

- $h^2 = \| h_1 \|_2^2 + \| h_2 \|_2^2 = \| h_{21} \|^2 + \| h_{22} \|^2$ is the power gain of the channel.
- I_2 is the 2×2 identity matrix.
- $S = [S_0, S_1]^T$ represents the transmitted symbols.
- $\tilde{n} = \begin{pmatrix} h_{11}^* n_0(1) + h_{12}^* n_0(2) + h_{21} n_1^*(1) + h_{22} n_1^*(2) \\ h_{21}^* n_0(1) + h_{22}^* n_0(2) - h_{11} n_1^*(1) + h_{12} n_1^*(2) \end{pmatrix}$ is some modified noise.

In order to take the channel correlation into account, which has a strong impact on the achievable performance of the system, different spatial channel models are considered.

3.3.4.4 Channel Capacity of MIMO System

Channel capacity of MIMO system with reference to Eqs. (3.6) and (3.7) for SIMO and MISO respectively, the channel capacity of MIMO system is given by:

$$C_{MIMO} = B \cdot \log_2 \left[1 + M \times N \times SNR \right] (bps / Hz) \tag{3.22}$$

Here M is no. of transmitting antennas and N is no. of receiving antennas. MIMO offers highest capacity among all diversity techniques.

3.3.4.5 Various Channel Models for MIMO System

Here various channels have been discussed which can be part of the MIMO wireless transmission. They are as under as per their characteristics.

- **Narrowband vs. Wideband**: If the channel coherent bandwidth is larger than the baseband signal, then the channel is called a narrowband channel. It is also called a flat channel because each transmitted frequency component undergoes the same fading. The radio channel is considered as wideband channel when the channel coherent bandwidth is less than the baseband signals. It is sometimes called frequency selective fading channel because each transmitted frequency component undergoes different fading. The channel medium is very dispersive in a frequency selective fading channel. In this environment, the received signal suffers from a delayed, distorted, and attenuated version of the transmitted signal. This generates intersymbol interference (ISI) which results in decrement in channel performance. In MIMO also two types of channel can be introduced. First is the wideband channel model which takes propagation channel as frequency selective which gives different response to various frequency band of transmitted bandwidth. Other channel may assume as the narrowband models which is flat fading channel and therefore the channel has the same response over the entire system bandwidth.

- **Physical vs. Non-physical Models**: The MIMO channel models can also be divided into physical and non-physical models. Some typical parameters are identified in this channel which include an angle of arrival, the angle of departure, and time of arrival. However, under many propagation conditions, the MIMO channels are not well described by a small set of physical parameters, and this limitation makes difficult to identify and validate the models. Another kind is non-physical models. They are based on the channel statistical characteristics. In general, the non-physical models are easy to simulate under which they were identified as they provide limited propagation characteristics, such as the bandwidth, configuration, and aperture of the arrays, and the heights of the transmit and receive antennas in the measurements.

- **Measurement vs. Scattering Models**: Another crucial approach to design MIMO channel is to measure the real MIMO channel responses through real field measurements. Some important parameters through the reading and measurement of the MIMO channel can be listed from recorded data, and the MIMO channel can be modeled to have similar statistical characteristics. A different approach is to assume a model that attempts to capture the channel characteristics. Such a model can often demonstrate the important characteristics of the MIMO channel as long as the constructed scattering environment is acceptable.

3.3.4.6   Advantages and Disadvantages of MIMO System

In this section, the merits and demerits of the MIMO system are described.

- **Advantages of MIMO System**: Any wireless system is designed with motive to increase throughput and transmission quality. MIMO systems can take multipath propagation positively and signal can be strengthened. Multipath random fading and multipath delay spread can be used to increase throughput. BER can be improved in MIMO systems without the need to increase bandwidth and/or power. MIMO improves throughput as well as transmission quality. Diversity is a technology used in MIMO for this purpose. Multiple antennas can be used to minimize the effect of fading caused by multipath propagation. When the antennas at the receive side are adequately spaced, then several copies of the transmitted signal are received through different channels and with different fading [4]. Because of multipath propagation, all the components cannot be affected by deep fading equally. Thus, diversity can improve signal quality.
- **One obvious disadvantage of MIMO is that they contain more antennas**: MIMO increases complexity, volume, and hardware costs of the system compared to SISO. MIMO systems are not always beneficial knowing that channel conditions depend on the radio environment. When there is Line of Sight (LOS), a higher LOS strength at receive will result in better performance and capacity in SISO system, while in MIMO systems capacity is reduced with higher LOS strength. This is beacuse strong contributions from LOS lead to higher correlation among antennas, which reduce the advantage of using a MIMO system.

By considering the all above facts, the following section of the chapter describes the real-time implementation of antenna diversity algorithms with Alamouti coding scheme so as to realize the system performance in the presence of diversities by analyzing relationships between BERs and SNRs.

3.4   Designing of Antenna Diversity Systems with Alamouti Coding Scheme

The practical implementation of various antenna diversity systems is discussed in this section. The performance of systems is evaluated using parameters like a signal to noise ratio (SNR) and bit error rate (BER) based on user-defined values. Figure 3.13 shows the designing flow of a wireless communication system with these antenna diversity systems. This system performs and contains various types of subsystems such as synchronization of data, estimation of the channel, decoding of source and channel. The implementation of this system is done with the help of MATLAB software where the simulation results of the system are generated and based on it, various curves of BER/SNR are plotted.

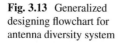

Fig. 3.13 Generalized designing flowchart for antenna diversity system

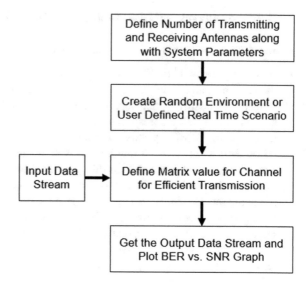

3.4.1 Implementation of SISO Antenna Diversity System

The implementation steps for SISO antenna diversity system is shown in Fig. 3.14. For the implementation of this system, various parameters such as no. of transmitting and receiving antennas, input values of data, type of modulation technique, the order of modulation, and signal to noise ratio need to set as per system requirement. Here, the M-ary PSK method and Rayleigh are used as a modulation method and channel, respectively. The values of signal, noise, and channel are generated in terms of a matrix with the help of MATLAB. For easy understanding of system working, a small number of user-defined values as input data are used. The same implementation can be used for a large amount of data.

The development of a system is initiated with the help of MATLAB as per the description given in Fig. 3.14. Here, ten user-defined symbols as input data are taken for checking of errors produced at the receiver side due to the wireless communication system. The sample data like 10 symbols or 20 bits are transmitted using SISO antenna system with single-channel matrix and single noise matrix. These numerical values are summarized in Table 3.1. With no diversity method, the same number of symbols are received by receiver with some errors. For analysis purpose, the channel SNR value is set to fix as 0.75 dB.

During transmission, the channel generates noise in the input symbols/bits which has been shown by red colored font. Under no diversity scenario, the receiver receives four correct symbols out of ten transmitted symbols. This indicated that six symbols are corrupted due to the multipath property of the wireless communication

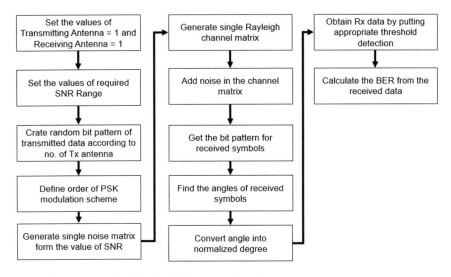

Fig. 3.14 Designing algorithm for SISO antenna diversity system

Table 3.1 Simulation results of SISO antenna system

Transmission symbols	2	1	1	1	0	0	2	1	3	2	Total no. of symbols: 10
Received symbols	1	2	3	2	3	0	2	2	3	2	Error in received symbols: 6
Matrix value for channel	\-0.151110462752079 + 0.0598519466936464*i*										Addition of signal matrix and noise matrix
Transmission bits	1	0	0	0	0	0	1	0	1	1	Total no. of bits: 20
	0	1	1	1	0	0	0	1	1	0	
Received bits	0	1	1	1	1	0	1	1	1	1	Error in received bits: 11
	1	0	1	0	1	0	0	0	1	0	
SNR	0.75 dB										Fix
Symbols to error rate (SER)	0.6 dB										
Bit error rate (BER)	0.55 dB										

system. It is also indicated that BER is less than SER which happens due to each symbol has two bits for representation and it might be the case that one bit is corrupted out of two bits. Now, this system is simulated with the help of 10,000 input bits and obtained BER vs. SNR graph which is shown in Fig. 3.15. This graph gives real-time data analysis of the SISO system.

As can be seen from Fig. 3.15, in case of no diversity, i.e. using single transmitting and single receiving antenna, initially the BER can be obtained around 0.1 at lower SNR = 1dB and at higher value of SNR, i.e. at SNR = 21dB, the achievable BER decreases around 0.02. In fact throughput of the system totally depends on the channel SNR.

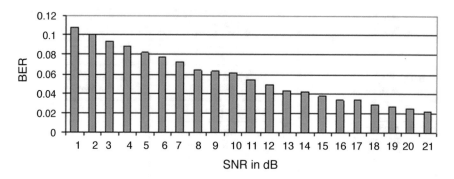

Fig. 3.15 BER vs. SNR for SISO antenna diversity system

3.4.2 Implementation of SIMO Antenna Diversity System

The implementation steps for SIMO antenna diversity system are shown in Fig. 3.16. For the implementation of this system, one transmitting antenna and two receiving antennas along with *M*-ary modulation method and Rayleigh channel are used. Here, two noise matrices along with single-channel matrix are generated with the help of MATLAB. At the receiver side, as per the value of modulation order, symbols are received by the receiver which may be converted in bits for proper analysis. Finally, the BER and SER are calculated as per received symbols and plot the graph between BER and SNR. In this system, one transmitting antenna and two receiving antennas are used. The simulation results of this system are summarized in Table 3.2. Here, 10 symbols are transmitted and 6 symbols got an error at the receiver side, which indicates that this system has SER = 0.6 dB, while 9 bits are corrupted out of 20 received bits, which indicates that this system has BER = 0.45 dB.

Now, this system has been implemented for 10,000 input bits and obtained BER value with different range of SNR. The simulation result for this scenario is realized in terms of BER vs. SNR graph which is shown in Fig. 3.17. This graph indicates that using this system, the value of BER is improved for a single communication channel. For the same input data, the SIMO system provides BER value of 0.45 dB, while the SISO system provides BER value of 0.55 dB. This situation indicates that this system provides better performance compared to SISO system.

3.4.3 Implementation of MISO Antenna Diversity System Along with Alamouti Coding

The implementation steps of MISO antenna diversity system along Alamouti coding are shown in Fig. 3.18. Here, two transmitting antennas and one receiving antenna along with random input symbols are taken for analysis of the system. The Alamouti coding is used where diversity of transmitter is taken place in the system.

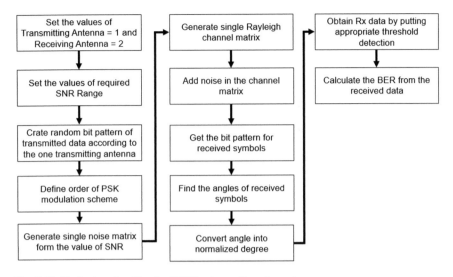

Fig. 3.16 Designing algorithm for SIMO antenna diversity system

Table 3.2 Simulation results of SIMO antenna system

Transmission symbols	2	1	1	1	0	0	2	1	3	2	Total no. of symbols: 10
Received symbols	1	2	1	1	1	3	0	1	3	0	Error in received symbols: 6
Matrix value for channel											Addition of channel matrix
Transmission bits	1	0	0	0	0	0	1	0	1	1	Total bits: 20
	0	1	1	1	0	0	0	1	1	0	
Received bits	0	1	0	0	0	1	0	0	1	0	Error bits: 9
	1	0	1	1	1	1	0	1	1	0	
SNR	0.75 dB										Fix
SER	0.6 dB										
BER	0.45 dB										

Matrix value for channel: $0.379772440973511 + 0.851240453939351i$ $0.578727415914267 - 0.0541236602159802i$ — Addition of channel matrix and two noise matrices

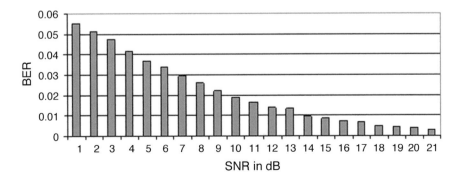

Fig. 3.17 BER vs. SNR for SIMO antenna diversity system

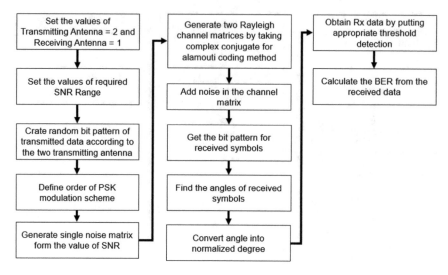

Fig. 3.18 Designing algorithm for MISO antenna diversity system with Alamouti coding

Table 3.3 Simulation results of MISO antenna diversity system with Alamouti coding

Transmission symbols	2	1	1	1	0	0	2	1	3	2	Total no. of
	2	1	1	1	0	0	2	1	3	2	symbols: 20
Received symbols	0	1	1	1	0	0	1	0	1	0	Error in received
	1	1	1	1	0	0	0	1	0	0	symbols: 9
Matrices values for channel	$-1.04429266452263 - 0.0993077287635684i\left(-S_1^*\right)$										Alamouti coding technique
	$-0.696253522655436 + 0.290063644892930i\ (S_0)$										
	$-0.696253522655436 - 0.290063644892930i\left(S_0^*\right)$										
	$1.04429266452263 - 0.0993077287635684i\ (S_1)$										
Transmission bits	1 0	0	0	0	0	0	1	0	1	1 0	Total bits: 40
		1	1	1	0	0	0	1	1		
	1 0	0	0	0	0	0	1	0	1	1 0	
		1	1	1	0	0	0	1	1		
Received bits	0	0	0	0	0	0	0	0	0	0	Error bits: 12
	0	1	1	1	0	0	1	0	1	0	
	0	0	0	0	0	0	0	0	0	0	
	1	1	1	1	0	0	0	1	0	0	
SNR	0.75 dB										Fix
SER	0.45 dB										
BER	0.30 dB										

Here, the complex conjugate of input symbols is taken and transmitted with the help of two antennas. For analysis purpose, signal matrix, two noise matrices which are complex conjugate, and two channel matrices are generated with the help of MATLAB. The simulation results of this system are summarized in Table 3.3.

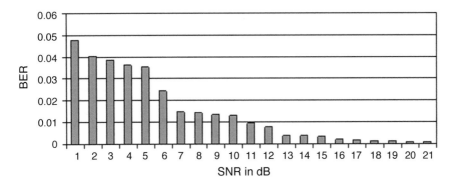

Fig. 3.19 BER vs. SNR for MISO antenna diversity system with Alamouti coding

In this MISO system, 20 symbols are transmitted with the help of two antennas along with two channel matrices and two noise matrices. The SER of this system is around 0.45 dB which indicated that 9 symbols are corrupted due to this system out of 20 input symbols, while BER of this system is 0.3 dB which indicated that 12 bits got corrupt out of 40 input bits. The BER vs. SNR of this system for 10,000 input bits is shown in Fig. 3.19. As per seen from this figure, the BER value is very low like 0.048 at a low value of SNR and decreases around 0.001 at a high value of SNR such as 21 dB. This indicated that this system performs better than the previous two systems such as SISO and SIMO.

3.4.4   Implementation of MIMO Antenna Diversity System Along with Alamouti Coding

The implementation steps of MIMO antenna diversity system along Alamouti coding are shown in Fig. 3.20. Here, two transmitting antennas and two receiving antennas along with random input symbols are taken for analysis of the system. Here, the complex conjugate of input symbols is taken and transmitted with the help of two antennas. For analysis purpose, two signal matrices, two noise matrices which are complex conjugate, and two channels matrices are generated with the help of MATLAB. The simulation results of this system are summarized in Table 3.4.

This system uses advantages of SIMO and MISO system to achieve good performance and improved BER by utilization of more transmitting antennas and receiving antennas. Here, two types of data such as 20 user-defined symbols or 40 bits and second applying the same phenomenon to stream of 10,000 bits for taking real time transmission analogy. As indicated in Table 3.4, only 8 received bits are corrupted out of 40 bits due to the use of 4 noise matrices and 2 channel matrices. This error rate is low compared to the other three systems. Also, BER = 0.2 and SER = 0.4 for this system is good compared to other diversity systems. The graph between BER

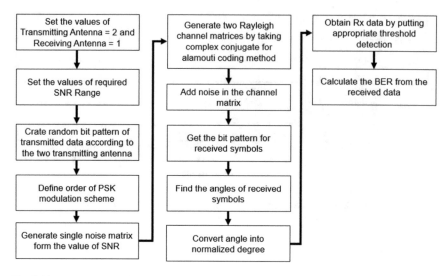

Fig. 3.20 Designing algorithm for MIMO antenna diversity system with Alamouti coding

Table 3.4 Simulation results of MIMO antenna diversity system with Alamouti coding

Transmission symbols	2	1	1	1	0	0	2	1	3	2		Total no. of symbols: 20
	2	1	1	1	0	0	2	1	3	2		
Received symbols	0	1	1	1	0	0	0	1	1	0		Error in received symbols: 8
	0	1	1	1	0	0	0	1	1	0		
Matrices values for channel	$1.09980602264884 - 0.852422471464323i\left(-S_1^*\right)$ $0.579729420304971 - 0.615471441657012i(S_0)$ $0.579729420304971 + 0.615471441657012i\left(S_0^*\right)$ $-1.09980602264884 - 0.852422471464323i(S_1)$ $0.0984368049147121 - 0.127313051276969i\left(-S_1^*\right)$ $0.637261486427597 - 0.481568123719595i(S_0)$ $0.637261486427597 + 0.481568123719595i\left(S_0^*\right)$ $0.0984368049147121 - 0.127313051276969i(S_1)$											Alamouti coding technique
Transmission bits	1 0	0 1	0 1	0 1	0 0	0 0	1 0	0 1	1 1	1 0		Total bits: 40
	1 0	0 1	0 1	0 1	0 0	0 0	1 0	0 1	1 1	1 0		
Received bits	0 0	0 1	0 1	0 1	0 0	0 0	0 0	0 1	0 1	0 0		Error bits: 8
	0 0	0 1	0 1	0 1	0 0	0 0	0 0	0 1	0 1	0 0		
SNR	0.75 dB											Fix
SER	0.4 dB											
BER	0.2 dB											

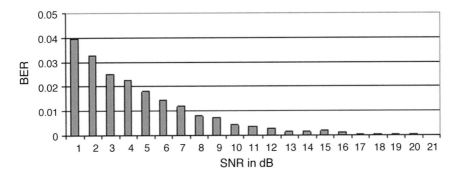

Fig. 3.21 BER vs. SNR for MIMO antenna diversity system with Alamouti coding

vs. SER for real-time data is shown in Fig. 3.21. In this graph, BER value is very low around 0.04 for the low value of SNR and near to 0 for the high value of SNR like 21 dB. This is indicated that reliable communication of data can be possible using this type of system.

3.4.5   Performance Comparison of Antenna Diversity Systems

This section gives a comparative analysis of various antenna diversity systems. The performance of systems is compared with the help of BER vs. SNR graph. Figure 3.22 shows a comparison of the performance of antenna diversity systems in terms of BER values. For example, at 5 dB of SNR, the BER of the SISO system is 0.08 while BER of SIMO is around 0.04. Similarly, the BER of MISO is around 0.035 for the same SNR value 0f 5 dB. The BER of MIMO is near 0.02 which is low along with all types of diversity system. This is indicated that MIMO system provides good performance compared to other diversity systems.

3.5   Designing of OFDM System

In this section, the basic information regarding of OFDM system along with its designing is discussed. This system provides a physical platform to build any WiMAX system. The basic model of a transceiver for OFDM along with its implementation is also discussed in this section.

3.5.1 Basic Concept of OFDM

The multi-carrier modulation (MCM) is the main idea behind OFDM, and OFDM obeys this concept. The basic of MCM is that it divides input bit stream into several parallel bit streams and they are used for modulation of subcarriers. The concept of OFDM is shown in Fig. 3.23. In this figure, each subcarrier is separated by guard band to avoid interference between them. In the receiver side, the bandpass filters are used to separate these parallel bit streams from the individual subcarriers. OFDM is a special version of MCM where subcarriers with orthogonal spaced and overlapping spectrums are used. Here, the bandpass filter is not used due to the orthogonality property of subcarriers.

The requirement of channel bandwidth can be reduced by the OFDM system which is clearly seen in Fig. 3.23. The orthogonality of input data streams can be achieved by applying FFT on it [5]. OFDM provides a high data rate for a large duration of symbol and eliminates the risk of intersymbol interference based on channel coherence time.

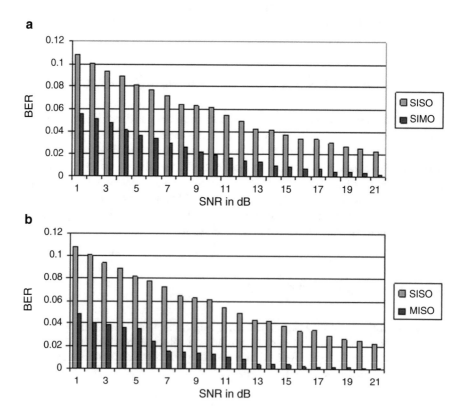

Fig. 3.22 Performance comparison of antenna diversity systems (**a**) between SISO and SIMO, (**b**) between SISO and MISO, (**c**) between SISO and MIMO, (**d**) all diversity systems

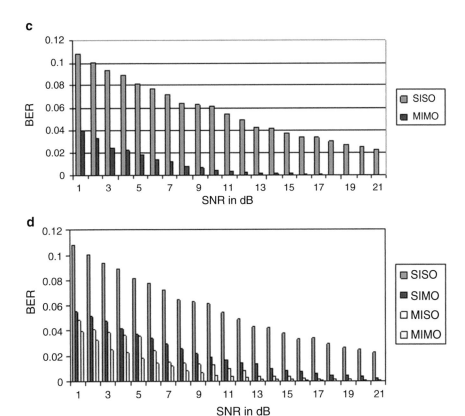

Fig. 3.22 (continued)

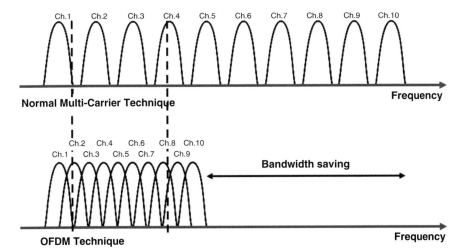

Fig. 3.23 Bandwidth saving in OFDM system

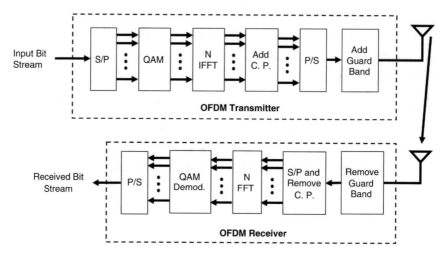

Fig. 3.24 Basic model for OFDM system

3.5.2 Working of OFDM System

The basic block diagram of the OFDM system is shown in Fig. 3.24. The OFDM transmitter comprises of basic blocks of QAM, IFFT and addition of cyclic prefix to evaluate the performance. The receiver possesses exactly opposite blocks of the system. Here, the single to parallel converter is used to convert single input data stream into the multi-dimensional stream. Then this stream gets modulated with the help of QAM and passes through the IFFT process. At the end of the transmitter side, the additional bit is added by cyclic prefix (CP) block to improve the performance of the system [5]. At the receiver side, the totally opposite process of transmitter is performed. After receiving that data, total opposite processes of cyclic prefix removal, FFT, QAM demodulation and parallel to serial conversion will take place.

3.5.3 Advantages, Disadvantages, and Applications of OFDM System

The OFDM system is mainly used for reducing the effect of ICI and combats the ISI. There are few advantages of this system which are discussed as per below:

- This system has high spectral efficiency due to overlapping spectrum.
- Simple implementation by fast Fourier Transform.
- The complexity of the receiver system is lower compared to the transmitter system.
- It can be used for transmission where the high data rate is required.

This system has some disadvantages which are discussed as per below:

- This system is very sensitive to high changes in times and frequency.
- This system used large no. of subcarriers which increase the power consumption of system compared to a single carrier system.

OFDM has gained a big interest since the beginning of the 1990s as many of the implementation difficulties have been overcome. OFDM has been in used or proposed for a number of wired and wireless applications. This is the first commercial users of this system. This system is also used for transmission of digital video signals [6]. After that, this system is accepted by worldwide researcher community and it is used as physical layer architecture for wireless LAN standards such as IEEE 802.11 a/g and IEEE 806.16 [7].

3.5.4 Designing of OFDM System

The main purpose of designing of OFDM system is to reduce data rate in subcarrier system. Hence, when the symbol rate increases then effects due to multipath are decreased. The insertion of higher valued CP will bring good results against combating multipath effects but at the same time it will increase loss of energy. Therefore, the tradeoff between these two parameters must be done before designing of OFDM system. The designing of the OFDM system requires some assumption which is discussed below:

- **Design Requirements for OFDM System**

  - **Bandwidth:** The channel bandwidth has a significant role in designing of OFDM system. The high bandwidth provides large no. of subcarriers which reduce the effect of the addition of CP.
  - **Bit Rate:** This is a no. of bit required as an input data stream.
  - **Delay:** The value of delay may be calculated by the length of CP and it is based on system requirement.
  - **Value of Doppler:** The effect of Doppler shifting must be taken into account for system designing.

The designing parameters required for OFDM system are discussed below:

- **Design Parameters for OFDM System**

  - **No. of Subcarriers:** The large no. of subcarriers reduces the effect of multipath but increases complexity in synchronization at the receiver side.
  - **Duration of Symbol and Length of CP:** The suitable value must be chosen which gives perfect ratio between symbol duration and length of CP.
  - **Spacing of Subcarrier:** The value of this parameter depends on the available channel bandwidth and no. of subcarriers.
  - **Type of Modulation:** The performance requirement will decide the selection of modulation scheme.

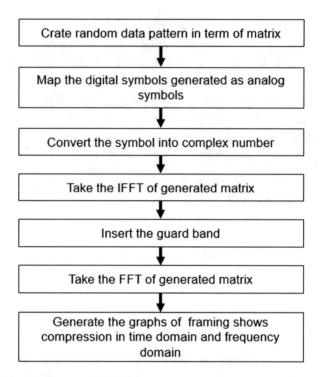

Fig. 3.25 Implementation steps for OFDM system

– **FEC Coding:** This coding provides robustness to the system against errors occurred in channels.

The implementation of OFDM system has been done with the help of MATLAB software. The steps for implementation of the OFDM system are shown in Fig. 3.25. For the initialization stage of implementation, first, set the value of symbol rate, symbol duration, no. of bit required for symbols, and guard band is selected. Based on these values, the input data stream is generated and after that, the process of the OFDM system is performed as shown in Fig. 3.24. The data stream is generated using "MODMAP" command in MATLAB and the orthogonal complex conjugate of the stream is generated using "AMODCE" command in MATLAB. By considering specific length of the symbols and length of the guard interval, IFFT of that stream is taken so as to convert the discrete frequency domain data into discrete time domain for making it compatible for real time transmission through the wireless channel.

Figure 3.26 shows the simulation results for the OFDM system with symbol per frame = 64 and 32, respectively. The time domain graphs in figures show orthogonality property of the system which requires less space and has low interference. It also indicated that due to orthogonality, the system requires less bandwidth for high

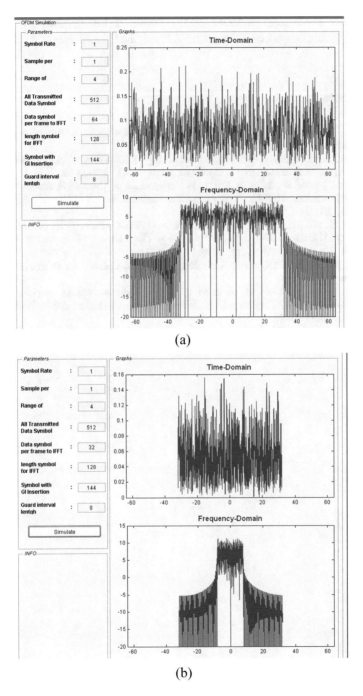

Fig. 3.26 Simulation results for OFDM system with (**a**) data symbol per frame = 64 and (**b**) data symbol per frame = 32

symbol rate. The system performance also indicated that it provides limited data rate, a significant amount of ISI and bandwidth saving for a different size of symbols.

References

1. Tse, D., & Viswanath, P. (2005). *Fundamentals of wireless communication*. Cambridge: Cambridge University Press.
2. Alamouti, S. M. (1998). A simple transmit diversity technique for wireless communications. *IEEE Journal on Selected Areas in Communications, 16*(8), 1451–1458.
3. Kansal, L., Kansal, A., & Singh, K. (2011). Performance of Alamouti Space-Time Coding in Fading Channels for IEEE 802.16 e protocol. *International Journal of Scientific & Engineering Research, 2*(7), 1–8.
4. Blum, R. S., Winters, J. H., & Sollenberger, N. R. (2001). On the capacity of cellular systems with MIMO. In *IEEE 54th Vehicular Technology Conference. VTC Fall 2001. Proceedings (Cat. No. 01CH37211)* (Vol. 2, pp. 1220–1224). IEEE.
5. Liu, H., & Li, G. (2005). *OFDM-based broadband wireless networks: Design and optimization*. New York: Wiley.
6. Jiang, T., Xiang, W., Chen, H. H., & Ni, Q. (2007). Multicast broadcast services support in OFDMA-based WiMAX systems [Advances in mobile multimedia]. *IEEE Communications Magazine, 45*(8), 78–86.
7. Teo, K. H., Tao, Z., & Zhang, J. (2007). The mobile broadband WiMAX standard [standards in a nutshell]. *IEEE Signal Processing Magazine, 24*(5), 144–148.

Chapter 4
WiMAX System Modeling

4.1 Modeling of WiMAX System Based on Antenna Diversity System and OFDM System

The complete physical modeling of the WiMAX system is discussed in the previous chapter. Figure 4.1 shows the modeling of the transmitter section and the receiver section of the WiMAX system. This system was the implementation with the help of antenna diversity technique and OFDM. The following points must be a consideration for designing this system.

- Set the values of system parameters such as order of modulation, cyclic prefix, no. of antennas for the transmitter as well as the receiver, OFDM symbols, and input data for the processing of the system.
- The random input data is generated using a random number generator and encoded using FEC coder for the secure transmission of data. After encoding the data, the block interleaver process is applied to it for reducing errors in it.
- OFDM transmitter transmits this modulated data by taking advantage of orthogonality to save bandwidth and reduce ISI effects.
- Now based on used diversity method, this data is transmitted over a wireless channel using various transmission methods such as different paths or channel estimation or Alamouti coding.
- At the receiver side, modulated and encoded data is given to OFDM demodulator and decoder to obtain original value of input random data.
- After obtaining received input random data, the distortion in this data is calculated with the help of bit error rate by comparing with original input data. Based on this comparison, the performance of the system will be judged.

The above-mentioned brief points must be thoroughly evaluated to design an efficient WiMAX system. The following subsections have discussed all steps in detail along with its specification. The designing and modeling flow for transmitter section and receiver section of the WiMAX system with a wireless channel is explained as follows.

© Springer Nature Switzerland AG 2020
B. S. Sedani et al., *WiMAX Modeling: Techniques and Applications*,
https://doi.org/10.1007/978-3-030-22460-8_4

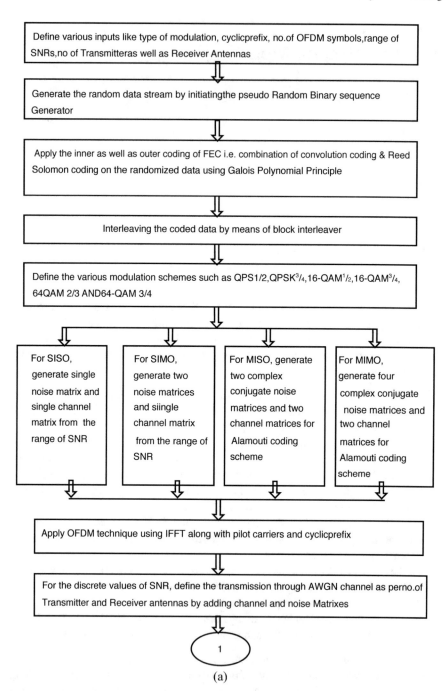

Fig. 4.1 Algorithm for WiMAX system model. (a) For WiMAX transmitter. (b) For WiMAX receiver

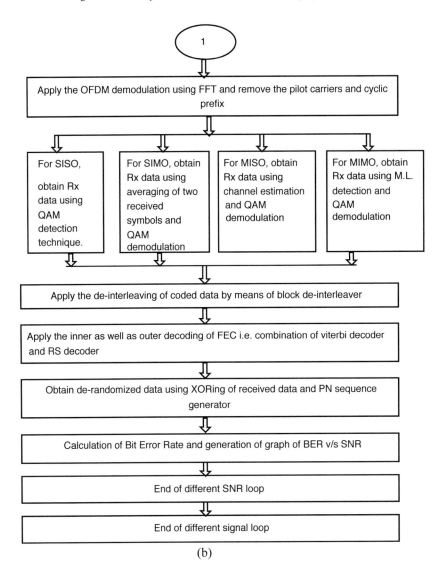

Fig. 4.1 (continued)

4.1.1   Modeling of WiMAX Transmitter

In this section, designing the flow of various subsystems of the WiMAX transmitter is explained with their design specifications.

(A) Data Initialization:

The steps for data initialization are as per below:

Step 1: Set the value of input parameters of the system such as no. of antennas for transmitter section and receiver section, the value of cyclic prefix, and rate value.

Step 2: Set the rate value for the selected modulation scheme.

Step 3: Set the cyclic prefix value for OFDM technique. The default value is 0.25, 0.125, 0.0625, and 0.03125.

Step 4: Set the symbol value for the OFDM technique.

Step 5: Finally, set the no. of antenna for transmitter and receiver as per chosen antenna diversity system.

(B) Data Generation and Randomization:

The steps for data generation are as per below:

Step 1: The input data is generated using a data generator. This data has a sequence of 1's and 0's. The size of this data depends on the rate value of the modulation scheme and it is a fixed value.

Step 2: The random stream of data is generated using a pseudo random sequence generator.

Step 3: After that, XOR process is performed between the output of data generator and the output of PN sequence generator for ciphering purpose.

Step 4: One byte of 0's is added after this process for modulation and encoding process.

Step 5: The padding of one byte is added to this random data after the encoding process.

(C) Reed–Solomon Encoding:

The steps for Reed–Solomon encoding are as per below:

Step 1: The encoding of input data is performed using the RS encoding method where different rate value is used for security purpose according to the modulation scheme. For this process, "RSENC" command is used in MATLAB.

Step 2: Here, two modes are defined for either transmission or reception such as "10" or "01." The value of TxRx is set "10" for encoding purpose.

Step 3: The value of codeword length (n) and data length (k) are also required for the encoding process. The value of these parameters is set according to the modulation scheme.

Step 4: For the implementation of Reed–Solomon encoding, the decimal number is required as input. Therefore, binary input data is converted into decimal using below polynomial functions such as primitive (p) and generator (g):

$$p(x) = x^8 + x^4 + x^3 + x^2 + 1 \tag{4.1}$$

$$g(x) = \left(x + \lambda^0\right)\left(x + \lambda^1\right)\left(x + \lambda^2\right)\ldots\left(x + \lambda^{2t-1}\right) \tag{4.2}$$

Step 5: The encoded decimal data is generated using a command like **CodeRS (msg, n, k)**.

Step 6:   Finally, the encoded decimal data is again converted into binary coded data for further process.

(D) Convolution Encoding:

The steps for convolution encoding are as per below:

Step 1:   After RS encoding, the convolution encoding is applied on RS encoded data for better performance and transparency.

Step 2:   Here, two modes are defined for either transmission or reception such as "10" or "01." The value of TxRx is set "10" for encoding purpose. Also, set parameters such as the rate for coding, bit values for generator polynomial, and constraint length according to the modulation scheme. For this parameter, **poly2trellis** command is used in MATLAB.

(E) Data Interleaving:

The steps for data interleaving are as per below:

Step 1:   Reordering of the encoded data is performed using different interleaving levels. Here, the value of interleaving level is set 12.

Step 2:   According to modulation scheme, no. of bits allocated to subcarrier for OFDM symbol (Ncbps) and no. of code per carrier (Ncpc) are set. Data interleaving is performed using below MATLAB commands:

```
k=0: Ncbps-1
mk = ((Ncbps/12) * mod(k,12)) + floor (k/12);
jk = s*floor(mk/s) + mod (mk + Ncbps-floor(12*mk/Ncbps), s);
```

Step 3:   After this process, the data is transmitted to modulation block for further process.

(F) M-QAM Modulation:

The steps for M-QAM modulation are as per below:

Step 1:   Set the value of modulation order M according to the modulation scheme. Here, modulation schemes such as QPSK and QAM with different order are used. The no. of bits per symbol (Xsym) is also defined according to the modulation scheme.

Step 2:   The following MATLAB commands are used to generate modulated data according to the modulation order and Xsym.

```
pskmod (xsym,M); for QPSK
qammod (xsym, M); for M-ary QAM
```

Step 3:   The output of this process is complex data which is based on the used modulation scheme.

Table 4.1 Default modeling parameters for various modulation schemes

Modulation scheme	Actual block of data (in bytes)	Encoded block of data (in bytes)	Coding rate	Sequence for RS coding	Convolution encoding
BPSK	12	24	1/2	(12,12,0)	1/2
QPSK	24	48	1/2	(34,24,4)	2/3
QPSK	36	48	3/4	(40,36,2)	5/6
16-QAM	48	96	1/2	(64,48,8)	2/3
16-QAM	72	96	3/4	(80,72,4)	5/6
64-QAM	96	144	2/3	(108,96,6)	3/4
64-QAM	108	144	3/4	(120,108,6)	5/6

In step no. D to F, various combinations of modulation technique along with specific coding technique according to the IEEE 802.16 standard can be implemented in the system to verify its performance in terms of variation in BER with respect to SNR. Table 4.1 shows the default parameters for various modulation schemes as per IEEE 802.16 standard [1–3].

(G) **Implementation of Diversity Schemes**:
The steps for implementation of diversity schemes are as per below:

Step 1: For SISO scheme, one data matrix for transmitting antenna and one noise matrix for receiving antenna are defined.

Step 2: For SIMO scheme, one data matrix for transmitting antenna and two noise matrices for two receiving antennas are defined.

Step 3: For the MISO scheme, two complex conjugate channels for two transmitting antennas and two noise matrices according to Alamouti coding for one receiving antenna are defined.

Step 4: For MIMO scheme, two complex conjugate channels for two transmitting antennas and four noise matrices according to Alamouti coding for two receiving antennas are defined.

(H) **OFDM Transmitter**:
The steps for OFDM transmitter are as per below:

Step 1: Set the value for transmission mode such as "10" and initialize the parameters for cyclic prefix, guard band, and points for FFT.

Step 2: The 256 points FFT based OFDM scheme for WiMAX system is designed using three types of data such as input data, pilot carrier, and subcarrier.

Step 3: Create OFDM symbols using three data types. Then take inverse FFT of this OFDM symbols using below command in MATLAB:

```
symbol_ofdm = sqrt (Nfft) . * ifft (symbol_ofdm, Nfft);
```

Step 4: Transmit the complete OFDM symbols through the wireless channel.

(I) Transmission through AWGN Channel:
The steps for transmission through AWGN channel are as per below:

Step 1:   Set the fixed value for SNR.
Step 2:   Generate channel matrix and noise matrix according to used diversity scheme for effective wireless communication.

4.1.2   Modeling of WiMAX Receiver

In this section, the designing flow of various subsystems of the WiMAX receiver is explained with their design specifications.

(A) OFDM Receiver:
The steps for OFDM receiver are as per below:

Step 1:   Set reception by setting value "01" and get transmit OFDM symbol. After that, the cyclic prefix is removed from received data which are added during transmission.
Step 2:   Perform the inverse of IFFT (FFT) to convert symbol into frequency domain by using the below command to get modulated data:

```
symbol_rx = fft (symbol_ofdm_rx, Nfft). / sqrt (Nfft);
```

Step 3:   After this process, the data are estimated using position information of the pilot carrier and subcarrier. This estimated data is sent to demodulation block for further process.

(B) M-QAM Demodulation:
The steps for M-QAM demodulation are as per below:

Step 1:   Set the value of modulation order M according to the modulation scheme.
Step 2:   The following MATLAB commands are used to get demodulated data according to the modulation order and Xsym.

```
zsym = pskdemod (yrx, M); QPSK
zsym = qamdemod (yrx, M); for M-ary QAM
```

Step 3:   Perform symbol to bet mapping to retrieve the original coded bits.

(C) Data De-interleaving:
The steps for data de-interleaving are as per below:

Step 1:   This is a reverse process of data interleaving.

Step 2: According to modulation scheme, no. of bits allocated to subcarrier for
         OFDM symbol (Ncbps) and no. of code per carrier (Ncpc) are set. Data
         de-interleaving is performed using below MATLAB commands:

```
j = 0: Ncbps-1
mj = s*floor(j/s) + mod ((j + floor(12*j/Ncbps)), s);
kj = 12*mj-(Ncbps-1) *floor(12*mj/Ncbps);
```

Step 3: Finally, the data is rearranged and given to convolution decoder block for
         further process.

(D) Convolution Decoding:
The steps for convolution decoding are as per below:

Step 1: For decoding purpose, set TxRx = "01," and take the value of parameters
         such as the rate for coding, bit values for generator polynomial, and con-
         straint length according to a modulation scheme. The decoding of data is
         performed using below the instruction:

```
decoded_data = vitdec (data_in, t,12);
```

Step 2: Remove the puncturing vector. Apply the decoded data to next stage of RS
         decoding.

(E) Reed–Solomon Decoding:
The steps for Reed–Solomon decoding are as per below:

Step 1: This is the reverse process of RS encoding. Convert decoded data into a
         decimal number for the proper working of RS decoder.
Step 2: For decoding purpose, set TxRx = "01," and the value of codeword length
         (n) and data length (k) are also required for the decoding process. The val-
         ues of these parameters are the same as generating at the transmitter
         section.
Step 3: The decoding of the data is performed using the below instruction:

```
decodeRS = rsdec (msg, n, k);
```

Step 4: Finally, the decoded decimal data is converted into binary data.

(F) Data De-randomization:
The steps for data de-randomization are as per below:

Step 1: This process removes padded one byte of zero bits from the decoded binary
         data.
Step 2: After that, XORing between random data and decoded binary data is per-
         formed to get actual bits of received data at the receiver side.

(G) **BER Calculation**:

The steps for BER calculation are as per below:

Step 1:   The decoded data stream will be compared with original data stream of the transmitter side and Bit Error Rate will be calculated using MATLAB function 'BITERR'.

Step 2:   Based on the calculated BER, the graph between SNR and BER will be the plot for the evaluation of system performance.

After discussion on complete designing and modeling of the WiMAX system, the MATLAB platform is used for testing of this model. Figure 4.2 shows the initialization of simulation parameters for the WiMAX system.

The default parameter values for the WiMAX system model are summarized in Table 4.2. These parameters are set as default for all antenna diversity schemes for better performance comparison of system.

```
***Select Modulation Order, Cyclic Prefix, Number of Symbol, Number of Tx and Rx
Antenna.
 --------------------------------------------------------------------------------
|Modulation Order| QPSK 1/2 | QPSK 3/4 | 16-QAM 1/2 | 16-QAM 3/4 | 64-QAM 2/3 | 64QAM-3/4 |
|----------------|----------|----------|------------|------------|------------|----------|
|rate id         |    1     |    2     |     3      |     4      |     5      |    6     |
 --------------------------------------------------------------------------------

       Enter the Modulation Order M =

       Enter the value of Cyclic Prefix [1/4 1/8 1/16 1/32] G=

       Enter the number of symbol =

       Enter the No of Tx Antenna =

       Enter the No of Rx Antenna =
```

Fig. 4.2   Initialization for modeling of WiMAX system

Table 4.2   Default parameters for modeling of WiMAX system

Parameters	Value
Cyclic prefix	1/16
OFDM symbols	256
Modulation order	16 QAM 3/4 (M = 4)
Parameters of RS encoder	(80,72,4)
The rate of the convolution encoder	5/6
No. of transmitting antenna	According to antennae diversity scheme
No. of receiving antenna	According to antennae diversity scheme

4.2   Designing and Analysis of WiMAX System Model Based on SISO Antenna Diversity System

The modeling of the WiMAX system for SISO antenna diversity scheme is presented along with its performance in this section. Figure 4.3 shows the initialization of the WiMAX system model for SISO antenna diversity scheme. The value of SNR would be selected in such a way that valid BER value can be obtained for the designed system.

After the set value of parameters for a WiMAX system model based on SISO antenna diversity scheme, the performance of the system in terms of the graph between BER versus SNR is shown in Fig. 4.4. Figure 4.4 shows that BER value is high for low SNR value but BER value is decreased when SNR value is increased. This figure also shows that the performance of this system is good when the SNR value is 1 dB.

```
***Select Modulation Order, Cyclic Prefix, Number of Symbol, Number of Tx and Rx
Antenna.
  ------------------------------------------------------------------------------------
|Modulation Order| QPSK 1/2 | QPSK 3/4 | 16-QAM 1/2 | 16-QAM 3/4 | 64-QAM 2/3 | 64QAM-3/4 |
|----------------|----------|----------|------------|------------|------------|-----------|
|rate id         |    1     |    2     |     3      |     4      |     5      |     6     |
  ------------------------------------------------------------------------------------

      Enter the Modulation Order M =   4

      Enter the value of Cyclic Prefix [1/4 1/8 1/16 1/32] G= 1/16

      Enter the number of symbol =  256

      Enter the No of Tx Antenna =  1

      Enter the No of Rx Antenna =  1
```

Fig. 4.3   Initialization of WiMAX system model based on SISO antenna diversity system

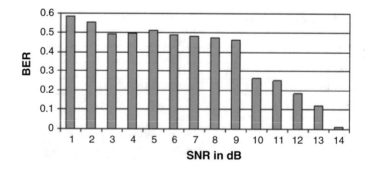

Fig. 4.4   BER vs. SER curve of WiMAX system model based on SISO antenna diversity system

4.3 Designing and Analysis of WiMAX System Model Based on SIMO Antenna Diversity System

The modeling of the WiMAX system for SIMO antenna diversity scheme is presented along with its performance in this section. Figure 4.5 shows the initialization of the WiMAX system model for SIMO antenna diversity scheme. The value of SNR would be selected in such a way that valid BER value can be obtained for the designed system.

After the set value of parameters for a WiMAX system model based on SIMO antenna diversity scheme, the performance of the system in terms of the graph between BER versus SNR is shown in Fig. 4.6. Figure 4.6 shows that BER value is high for low SNR value but BER value is decreased when SNR value is increased. This figure also shows that the performance of this system is good when the SNR value is 1 dB and shows that BER value is near to zero or zero after the SNR value is 9 dB or more.

```
***Select Modulation Order, Cyclic Prefix, Number of Symbol, Number of Tx and Rx
Antenna.
-------------------------------------------------------------------------------
|Modulation Order| QPSK 1/2 | QPSK 3/4 | 16-QAM 1/2 | 16-QAM 3/4 | 64-QAM 2/3 | 64QAM-3/4 |
|----------------|----------|----------|------------|------------|------------|-----------|
|rate id         |    1     |    2     |     3      |     4      |     5      |     6     |
-------------------------------------------------------------------------------

        Enter the Modulation Order M =   4

        Enter the value of Cyclic Prefix [1/4 1/8 1/16 1/32] G= 1/32

        Enter the number of symbol =  256

        Enter the No of Tx Antenna =  1

        Enter the No of Rx Antenna =  2
```

Fig. 4.5 Initialization of WiMAX system model based on SIMO antenna diversity system

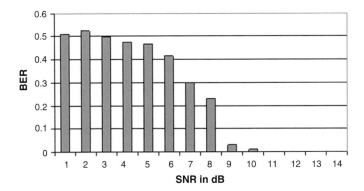

Fig. 4.6 BER vs. SER curve of WiMAX system model based on SIMO antenna diversity system

4.4 Designing and Analysis of WiMAX System Model Based on MISO Antenna Diversity System

The modeling of the WiMAX system for MISO antenna diversity scheme is presented along with its performance in this section. Figure 4.7 shows the initialization of the WiMAX system model for MISO antenna diversity scheme. The value of SNR would be selected in such a way that valid BER value can be obtained for the designed system.

After the set value of parameters for a WiMAX system model based on MISO antenna diversity scheme, the performance of the system in terms of the graph between BER versus SNR is shown in Fig. 4.8. Figure 4.8 shows that BER value is high for low SNR value but BER value is decreased when SNR value is increased. This figure also shows that the performance of this system is good when the SNR value is 1 dB and shows that BER value is near to zero or zero after the SNR value is 9 dB or more.

```
***Select Modulation Order, Cyclic Prefix, Number of Symbol, Number of Tx and Rx
Antenna.
    ---------------------------------------------------------------------------------
    |Modulation Order| QPSK 1/2 | QPSK 3/4 | 16-QAM 1/2 | 16-QAM 3/4 | 64-QAM 2/3 | 64QAM-3/4 |
    |----------------|----------|----------|------------|------------|------------|-----------|
    |rate id         |    1     |    2     |     3      |     4      |     5      |     6     |
    ---------------------------------------------------------------------------------

        Enter the Modulation Order M =    4

        Enter the value of Cyclic Prefix [1/4 1/8 1/16 1/32] G= 1/16

        Enter the number of symbol =  256

        Enter the No of Tx Antenna =  2

        Enter the No of Rx Antenna =  1
```

Fig. 4.7 Initialization of WiMAX system model based on MISO antenna diversity system

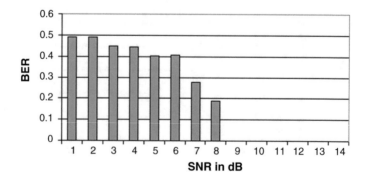

Fig. 4.8 BER vs. SER curve of WiMAX system model based on MISO antenna diversity system

4.5 Designing and Analysis of WiMAX System Model Based on MIMO Antenna Diversity System

The modeling of the WiMAX system for MIMO antenna diversity scheme is presented along with its performance in this section. Figure 4.9 shows the initialization of the WiMAX system model for MIMO antenna diversity scheme. The value of SNR would be selected in such a way that valid BER value can be obtained for the designed system.

After the set value of parameters for a WiMAX system model based on the MIMO antenna diversity scheme, the performance of the system in terms of the graph between BER versus SNR is shown in Fig. 4.10. Figure 4.10 shows that BER value is high for low SNR value but BER value is decreased when SNR value is increased. This figure also shows that the performance of this system is good when the SNR value is 1 dB and shows that BER value is near to zero or zero after the SNR value is 7 dB or more.

```
***Select Modulation Order, Cyclic Prefix, Number of Symbol, Number of Tx and Rx
Antenna.
 ------------------------------------------------------------------------------------
|Modulation Order| QPSK 1/2 | QPSK 3/4 | 16-QAM 1/2 | 16-QAM 3/4 | 64-QAM 2/3 | 64QAM-3/4 |
|----------------|----------|----------|------------|------------|------------|-----------|
|rate id         |    1     |    2     |     3      |     4      |     5      |     6     |
 ------------------------------------------------------------------------------------

        Enter the Modulation Order M =    4

        Enter the value of Cyclic Prefix [1/4 1/8 1/16 1/32] G= 1/16

        Enter the number of symbol =   256

        Enter the No of Tx Antenna =   2

        Enter the No of Rx Antenna =   2
```

Fig. 4.9 Initialization of WiMAX system model based on MIMO antenna diversity system

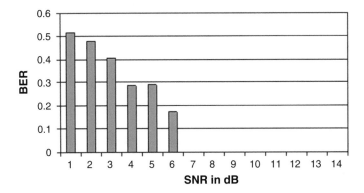

Fig. 4.10 BER vs. SER curve of WiMAX system model based on MIMO antenna diversity system

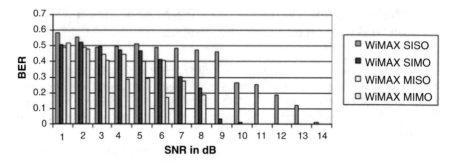

Fig. 4.11 Comparative comparison of performance for WiMAX system model based on all antenna diversity systems

4.6 Comparison of Performance for WiMAX System Model Based on All Antenna Diversity Systems

Sections 4.2–4.5 give the performance of a WiMAX system model for different antenna diversity schemes for various values of SNR. Here, the comparative analysis of different antenna diversity schemes for WiMAX system with parameter values such as cyclic prefix = 0.0625, 256 OFDM symbols, and M = 16 QAM 3/4 is given. Figure 4.11 shows the comparative analysis of modeling a WiMAX system for different antenna diversity schemes.

The comparison shows that the BER value for each scheme is obtained in the range of SNR values from 1 to 14 dB. The performance SISO scheme is very bad compared to other schemes in terms of obtained BER values. The performance of the MIMO scheme is best compared to other schemes. The highest BER values can be achieved using the SISO scheme.

The WiMAX system is capable of providing very huge coverage area of approximately 50 Km. But when the distance of communication increases, the performance of this system is decreased. Under such environment, implementation of MIMO is opening the great voyage towards highly efficient transmission at lower values of SNRs in WiMAX system. As per the presented work, just 6 dB SNR is required to reduce the errors in the received signal.

References

1. Eklund, C., Marks, R. B., Stanwood, K. L., & Wang, S. (2002). IEEE standard 802.16: A technical overview of the Wireless MAN™ air interface for broadband wireless access. *IEEE Communications Magazine, 40*(6), 98–107.
2. Khan, M. N., & Ghauri, S. (2008). The WiMAX 802.16 e physical layer model. In *IET International Conference on Wireless, Mobile and Multimedia Networks* (pp. 117–120).
3. Alim, O. A., Abdallah, H. S., & Elaskary, A. M. (2008). Simulation of WiMAX systems. In *IEEE Communication Workshop* (pp. 11–16).

Chapter 5
WiMAX System Modeling for Real-Time Data Transmission

5.1 Basic WiMAX System Model with System Parameters

In the previous chapter, the modeling of WiMAX system for different antenna diversity schemes is discussed using random binary data. The basic modeling of traditional WiMAX system along with detailing of every block has been already discussed in the Chap. 2. The ideal AWGN channel with no fading effects is used in this model. Table 5.1 summarized the default parameters used for modeling of WiMAX system based on different antenna diversity schemes.

Table 5.1 indicated that the parameters such as cyclic prefix and modulation order are treated as important parameters. As such these parameters can have multiple no. of values but after certain experimentations that have been discussed in Chap. 4, their optimum value can be derived. As further, this chapter discussed the transmission of digital image and speech signal using a WiMAX system. Figures 5.1 and 5.2 show the sample test image with 256 × 256 pixels and a sample speech signal with 2 s, respectively. Figure 5.3 shows the QAM symbols for transmission of original data.

5.2 Modeling of WiMAX-MIMO System for Real-Time Image Transmission

The modeling of WiMAX-MIMO system for transmission of the digital image is discussed in this section. Figure 5.4 shows the WiMAX-MIMO system for image transmission. A brief working of this model is explained in the following hierarchy.

1. An input image (256 × 256) is taken and converted into binary bits and added addition overhead bits (280 × 1) for proper processing of the system.

© Springer Nature Switzerland AG 2020
B. S. Sedani et al., *WiMAX Modeling: Techniques and Applications*,
https://doi.org/10.1007/978-3-030-22460-8_5

Table 5.1 Parameters for WiMAX system model

Model parameter	WiMAX-SISO system	WiMAX-SIMO system	WiMAX-MISO system	WiMAX-MIMO system
Model setup	MATLAB	MATLAB	MATLAB	MATLAB
SNR value of AWGN channel (dB)	24	24	24	24
Power rating of AWGN channel (W)	0.1	0.1	0.1	0.1
Image matrix size	256 × 256	256 × 256	256 × 256	256 × 256
Samples/frame	35	35	35	35
Generator polynomial of PN sequence generator	[1 0 0 0 0 0 0 0 0 0 0 0 0 0 1 1]	[1 0 0 0 0 0 0 0 0 0 0 0 0 0 1 1]	[1 0 0 0 0 0 0 0 0 0 0 0 0 0 1 1]	[1 0 0 0 0 0 0 0 0 0 0 0 0 0 1 1]
Initial states of PN sequence generator	[0 0 0 1 1 1 0 1 1 1 1 0 0 0 1]	[0 0 0 1 1 1 0 1 1 1 1 0 0 0 1]	[0 0 0 1 1 1 0 1 1 1 1 0 0 0 1]	[0 0 0 1 1 1 0 1 1 1 1 0 0 0 1]
Codeword length N of RS encoder	255	255	255	255
Message length K of RS encoder	239	239	239	239
RS encoder rate	3/4	3/4	3/4	3/4
Convolution encoder input vector length k	171	171	171	171
Convolution encoder output vector length n	131	131	131	131
Native rate of convolution coder	5/6	5/6	5/6	5/6
Puncture vector	([1 0 1 0 1;1 1 0 1 0], 10, 1)	([1 0 1 0 1;1 1 0 1 0], 10, 1)	([1 0 1 0 1;1 1 0 1 0], 10, 1)	([1 0 1 0 1;1 1 0 1 0], 10, 1)

QAM modulation order	4	4	4	4
QAM normalization factor	$1/\sqrt{2}$	$1/\sqrt{2}$	$1/\sqrt{2}$	$1/\sqrt{2}$
OFDM symbol time	64 µs	64 µs	64 µs	64 µs
Channel bandwidth	1.5–20 MHz	1.5–20 MHz	1.5–20 MHz	1.5–20 MHz
NFFT	256	256	256	256
No. of the cyclic prefix	1/8	1/8	1/8	1/8
OFDM training sequence	N.A.	N.A.	200 bits	200 bits
OFDM pilot carriers	55 bits	55 bits	55 bits	55 bits
No. of transmitter antenna for alamouti encoding	–	–	2	2
No. of receiver antenna for Alamouti decoding	–	–	1	2
Samples of the speech signal	16,000	16,000	16,000	16,000
Amount of noise generated by noise generators 1 and 2	AWGN_PWR*0.1*((randn (20000,1) + j*randn (20000,1))/ sqrt (AWGN_SNR))	AWGN_PWR*0.1*((randn (20000,1) + j*randn (20000,1))/ sqrt (AWGN_SNR))	AWGN_PWR*0.1*((randn (20000,1) + j*randn (20000,1))/ sqrt (AWGN_SNR))	AWGN_PWR*0.1*((randn (20000,1) + j*randn (20000,1))/ sqrt (AWGN_SNR))

Fig. 5.1 Test input image

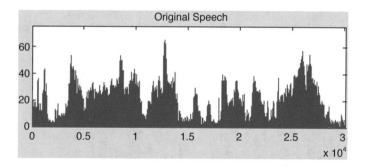

Fig. 5.2 Test speech signal

2. The random data (280×1) is generated using a randomizer and then encrypted using RS encoder and convolution encoder to get encrypted output data with a size of 384×1.
3. This encrypted data is modulated with the help of QAM modulator to get modulated data with a dimension of 192×1.
4. Here MIMO diversity scheme is used for transmission of data. Therefore, the modulated data is reframed to get original data with the dimension of 201×1.
5. After getting framed data, the main task of the system is performed using Alamouti encoder which is based on the logic of Alamouti coding. Figure 5.5 shows the basic model for Alamouti encoder. This model consists of OSTBC encoder and two U-Y selectors. Here, the logic of two-stage conversion is performed and the output of the model is data with dimension 201×2. Here, the output of Alamouti encoder is two data streams which are complex conjugation

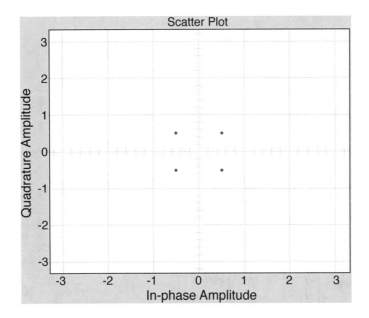

Fig. 5.3 QAM symbols for testing of performance of designed WiMAX system model

of each other. After that, these data are separated and sent using two OFDM transmitters.

6. Figure 5.6 shows the detailed model of the OFDM transmitter. Here, two transmitters worked on the same function of two different sequences by adding some guard bands and cyclic prefix in the data stream to reduce the ISI error during transmission. This process converts 3-D data in a 1-D matrix with a dimension of 864 × 1.

7. After that, two data streams with two different sequences are transmitted using two different channels. The nature of channels is AWGN and the basic structure of these channels is shown in Fig. 5.7. Here, the effect of multipath fading is not considered. The noise is added into data streams using two adders and the output of two channels is fed to two different OFDM receivers. If the first two adders are not used in the channels, then data can travel in different directions or paths and never combined which overrule the Alamouti logic. So that, by using the top two adders, the combination of data is performed and after that noise is added in the combined data using the other two adders.

8. The output of the channel is fed to two separate OFDM receivers. Figure 5.8 shows the basic structure of the OFDM receiver. The important functions of the receiver are to remove the guard bands and cyclic prefix.

9. The received data streams are fed to Alamouti decoder. Figure 5.9 shows the basic structure of this decoder which uses maximum likelihood concept for data decoding. It decoded original data with dimension 200 × 2 using the estimation of channel and STB combination. Here, for estimation of the channel, two training sequences are required along with the original data matrix which gives two-dimension output data.

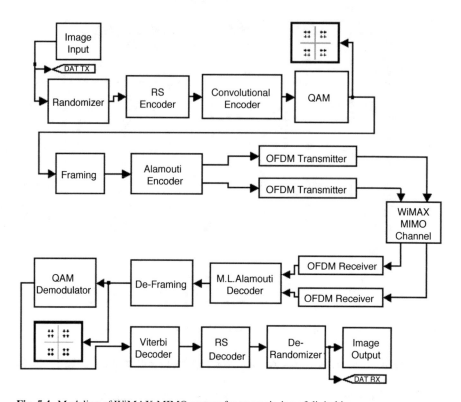

Fig. 5.4 Modeling of WiMAX-MIMO system for transmission of digital image

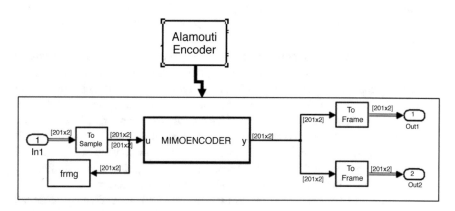

Fig. 5.5 Model for Alamouti encoder

10. After obtaining decoding data, the deframing of decoded data with dimension
    200 × 2 is performed using parallel to serial conversion and removes the
    additional zero bits from the decoded data to obtain original decoded data with
    dimension 192 × 1.

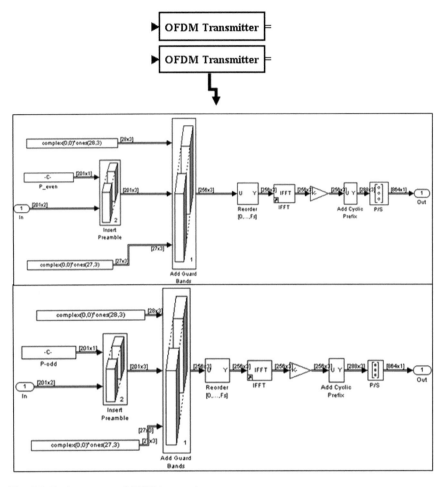

Fig. 5.6 Basic structure of OFDM transmitters

11. After this, viterbi decoding is applied to obtain original image at the receiver side.
12. At last, the comparison between output and input image is performed to calculate the Bit Error Rate.

Figure 5.11 shows the WiMAX-MIMO system after simulation. The model shows the input data values and output data values at each block with its justification. The variations in QAM symbols, received image, and BER value for system performance for transmission of the image are given in Figs. 5.12, 5.13, and 5.14, respectively.

The performance of the system shows that the number of bits for an image such as 635,880 is successfully transmitted using this model. This situation indicated that the BER of this model is very low around 0.00009 which fulfills the criteria of image data transmission.

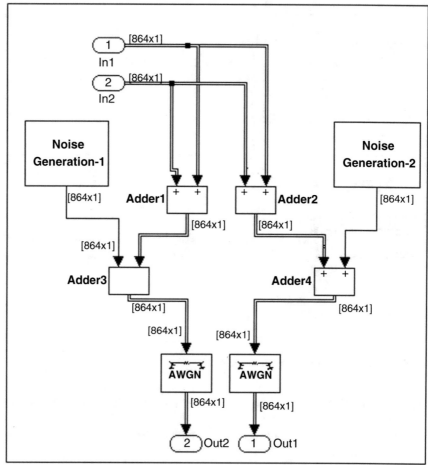

Fig. 5.7 Basic structure of AWGN channel

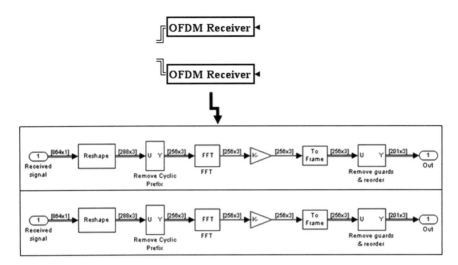

Fig. 5.8 Basic structure of OFDM receivers

5.3   Modeling of WiMAX-MIMO System for Real-Time Speech Signal Transmission

In this section, modeling of the WiMAX system based on the MIMO antenna diversity system for real-time speech signal transmission is discussed. The steps for transmission of the real-time speech signal are the same as discussed for transmission of the image, only changes taken place in input data and output data. The simulated WiMAX-MIMO system model for real-time speech signal transmission is shown in Fig. 5.15.

The variations in QAM symbols, received speech signal, and BER value for system performance for speech signal transmission are given in Figs. 5.16, 5.17, and 5.18, respectively. The performance of the system shows that the number of speech signal bits such as 280,280 are successfully transmitted using this model. This situation indicated that the BER of this model is very low around 0.0004 which fulfills the criteria of signal transmission.

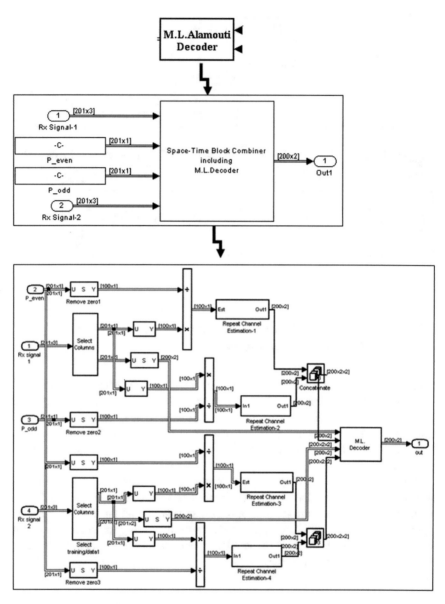

Fig. 5.9 Alamouti decoder for WiMAX-MIMO system

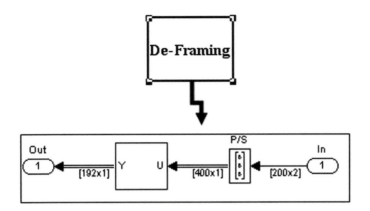

Fig. 5.10 Deframing for WiMAX-MIMO system

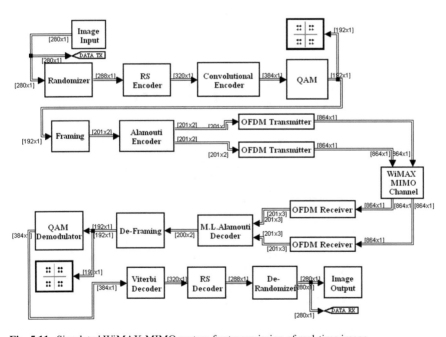

Fig. 5.11 Simulated WiMAX-MIMO system for transmission of real-time image

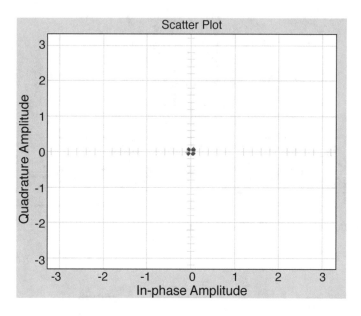

Fig. 5.12 Output QAM symbols for transmission of real-time image

Fig. 5.13 Received image

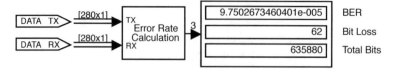

Fig. 5.14 BER calculator for transmission of real-time image

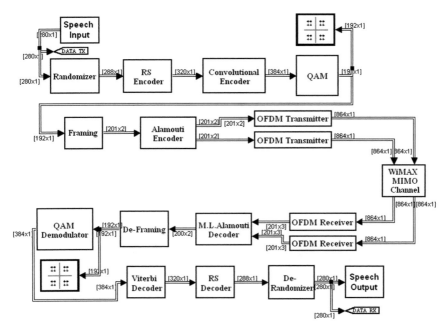

Fig. 5.15 Simulated WiMAX-MIMO system for real-time speech signal transmission

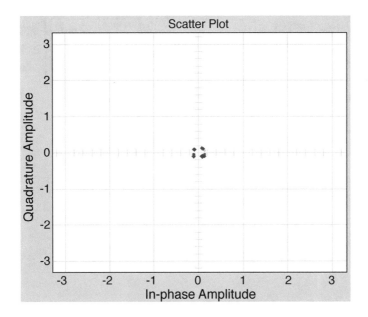

Fig. 5.16 Output QAM symbols for transmission of real-time speech signal

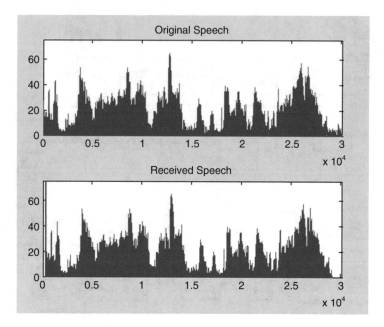

Fig. 5.17 Original and received speech signal

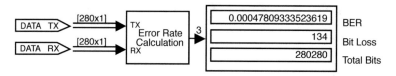

Fig. 5.18 BER calculator for transmission of real-time speech signal

Chapter 6
Summary of Book and Future Direction in WiMAX System Modeling

This chapter covers a summary of the book and future direction in WiMAX system modeling.

6.1 Important Points Highlighted in the Book

Wireless communication plays an important role in the long range transmission of data. One of the famous wireless standards such as WiMAX system has been used for effective communication in 4G mobile communication systems. The complete modeling of the WiMAX system according to IEEE 802.16 standard has been discussed with the help of MATLAB platform in this book. The performance of this designed system is judged by obtaining BER for specific SNR. For example, at SNR = 20 dB, the traditional WiMAX system possesses BER of 0.47 i.e. almost 50% bits are getting lost during the transmission. The same system gives BER of 0.004 at SNR = 27 dB, which indicates that the performance of WiMAX system is good for higher values of SNR. The multipath signal propagation and fading effects have affected the performance of WiMAX system. To combat this challenge, the channel SNR must be as high as possible which leads to the higher amount of signal power requirement against the existing noise level. But this solution is not feasible for an effective communication system.

This book is mainly developed for modeling of WiMAX system with different parameters such as coding rate, modulation order, and cyclic prefix. In this book, the model for the WiMAX system with the help of antenna diversity scheme along with Alamouti coding is discussed. The performance of different types of model for WiMAX system is analyzed and results show that MIMO antenna diversity scheme based WiMAX system is performed better than other antenna diversity schemes such as SISO, SIMO, and MISO based WiMAX systems. This book covers the real-time transmission of data using the WiMAX system. The designing model for this purpose is discussed with the help of digital image and speech signal. The results

© Springer Nature Switzerland AG 2020   113
B. S. Sedani et al., *WiMAX Modeling: Techniques and Applications*,
https://doi.org/10.1007/978-3-030-22460-8_6

show that this model is effectively used for transmission of digital images and speech signals. More specifically, research work inside this book evaluates the complete physical layer of WiMAX system along with the implementation of various antenna diversity techniques on the platform of MATLAB.

6.2 Future Research Direction in WiMAX Modeling

The main focus of WiMAX system modeling is to achieve the lowest BER i.e. to achieve high quality system performance. The performance of the system in terms of BER and capacity can be improved by increasing a greater number of antennas and through spatial multiplexing and transmission of signals through various which lead to the designing and implementation of BLAST structure in WiMAX frame work. Furthermore, the performance of the system will be evaluated and analyzed with the help of the OFDM system and different modulation schemes. For the same purpose, various models would be implemented with the help of DSP processor and the performance of these models would be analyzed.

Correction to: Introduction to WiMAX System

Correction to:
Chapter 1 in: B. S. Sedani et al.,
WiMAX Modeling: Techniques and Applications,
https://doi.org/10.1007/978-3-030-22460-8_1

The original version of this chapter was inadvertently published without text labels inside figures 1.1, 1.2, and 1.3. The correct figures with text labels are given below:

The updated online version of this chapter can be found at
https://doi.org/10.1007/978-3-030-22460-8_1

B. S. Sedani et al., *WiMAX Modeling: Techniques and Applications*,
https://doi.org/10.1007/978-3-030-22460-8_7

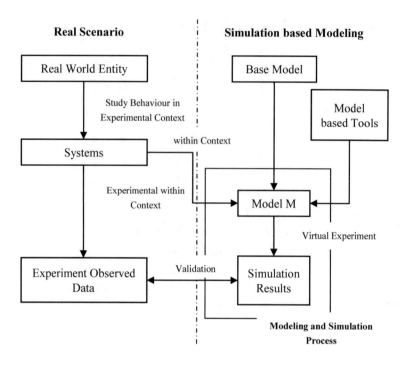

Fig. 1.1 Basic structure of system modeling

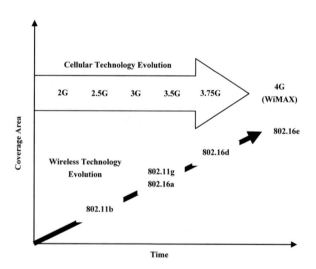

Fig. 1.2 Evolution of cellular and wireless technologies

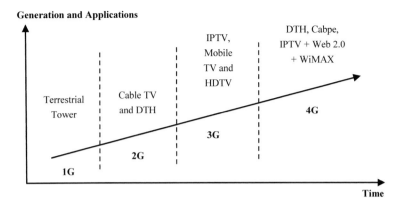

Fig. 1.3 Scenarios of wireless system generation

Index

© Springer Nature Switzerland AG 2020   115
B. S. Sedani et al., *WiMAX Modeling: Techniques and Applications*,
https://doi.org/10.1007/978-3-030-22460-8

Printed in the United States
by Baker & Taylor Publisher Services